Tuning the Mind

Tuning the Mind

Connecting Aesthetics to Cognitive Science

Ruth Katz and Ruth HaCohen

Transaction Publishers
New Brunswick (U.S.A.) and London (U.K.)

This book is printed on acid-free paper that meets the American National Standard for Permanence of Paper for Printed Library Materials.

Library of Congress Catalog Number: 2002072131
ISBN: 0-7658-0081-0
Printed in the United States of America

Library of Congress Cataloging-in-Publication Data

Katz, Ruth, 1927-
Tuning the mind : connecting aesthetics to cognitive science / Ruth Katz and Ruth HaCohen.
p. cm.
Includes bibliographical references and indexes.
ISBN 0-7658-0081-0 (alk. paper)
1. Music—Psychological aspects. 2. Cognition. 3. Music—Philosophy and aesthetics. I. HaCohen, Ruth. II. Title.

ML3838 .K28 2002
781'.11—dc21 2002072131

In Memory of Carl Dahlhaus (1928-1989)

Contents

Preface

This book grew out of an insight that the cognitive studies that have become so central to contemporary discourse owe an unacknowledged debt to eighteenth-century deliberations on how art, and especially music, is processed by the mind. We realized from the outset that to explore this intellectual turn would require an interdisciplinary effort to mobilize the entire cultural spectrum that affected these deliberations. We soon discovered that what came to fruition towards the end of the eighteenth century had started as far back as the end of the sixteenth century, and was not fully recognized by the initiators themselves. Indeed, it is only with hindsight, once modern art no longer needs justification and cognitive studies are thriving that the position of these deliberations in intellectual history can be properly assessed.

It began in the late Renaissance with the attempt to render art more expressive. In music, with which we are most familiar, it involved the quest for the lost formulae of the Greeks who allegedly knew how to use music to heighten the power of words. Whether real or feigned archaeologists, those engaged in the quest identified new declamatory forms for effective song speech, from which dramatic music and related forms have emerged. This, of course, was not rediscovery, but invention, from which a semantics of musical expression gradually sprouted. Over a period then, the systematic marriage of words and music branded music with meaning, while music endowed words with affective specificity. This had consequences for "pure" instrumental music.

What had taken place in music, we came to realize, had occurred in painting and poetry as well, although in less conspicuous ways. This is why music, in fact, became more and more emblematic for the arts in general, for it best exemplified what they all tried to achieve. It is explicated in the tale which narrates how *musica* replaced *pictura* in the Horatian paradigm of *ut pictura poesis.*

Whereas modern science in the early seventeenth century affected all of the arts, it revealed central epistemological difficulties in music that concern art in general. The new scientific discoveries, as is well known, displaced music from the honored place it had occupied in the *quadrivium*; this, however, was only symptomatic of a broader change that had taken place in the theory of knowledge. It is not that music's numerology was no longer believed to contain artistic truth, but that music was no longer considered able to disclose the secrets of *world* harmony. The Platonic conception which maintained that "a sameness of idea" pervaded all spheres of creation from the spiritual to the material and from the rational to the sensual, had become suspect. The recognition that only what is commensurable can be part of scientific theory, delegitimized old connections between the physical and the metaphysical, or in the case of music, between acoustic quantities and emotional qualities. At the same time, however, this skepticism brought about more careful differentiations between varied mental faculties. Thus, the crisis that could have devastated the arts, and especially music, gave rise to new insights concerning the intricate processing of mental qualities.

By the eighteenth century, writers could examine anew the role of the arts within a general theory of knowledge. This was made possible not only because the phenomena to be observed had by then come into being but also because new conceptions dealing with mental *in*commensurabilities had been refined. In this state of affairs, those equipped with the philosophical tools forged by both rationalists and empiricists, could treat the old aesthetic queries with new rigor.

Tracing this development, the first chapter of the book deals with the ascendance of epistemological queries concerning artistic qualities and tries to isolate their historical moment. We emphasize that questions were asked in all of the arts, and that they were equally related to the rise of expression as an artistic desideratum and to recognition that the immediacy achieved through artistic illusion is neither quantifiable, nor self-understood.

Awareness that the artist is the *creator* of artistic expression and the one who guarantees its immediacy emerged at the same time. This is the stage at which rhetoric joined poetics in presenting new ways of creating and of *understanding* the arts. In this new conceptual space Descartes's analysis of the habitual vs. innate elements in emotional behaviour looms large. Many later theories of art took the lead from Descartes. However, Marin Mersenne, Descartes's contemporary and friend, made the first modern attempt to connect artistic expression and metaphorical activation, and more specifically, musical movement with emotional behaviour.

Whereas the first chapter concentrates primarily on new aesthetic conceptions in the field of music, the second chapter attempts to show that in painting and literature as well, a new agenda was being formulated in the course of the seventeenth century. There was a decline in preoccupation with content—the ideas or objects represented by the work of art—which had assured its communicative value and standing until that time. Gradually it became clear that the power of art does not reside in the objects it represents or imitates, but rather in the creative modalities specific to each artistic medium, which influences its messages. The main part of this chapter explains how these new ideas connect with different modes of "musicalization."

The third chapter discusses the conditions which made music a paradigm for the other arts. Of course, from the point of view of mimesis, music was always problematic. In order to fathom its peculiar cognitive nature, it had to be connected, paradoxically, to meanings existing outside its own precincts. The primary and most important step in this direction involved an understanding of the potential for turning musical messages into "fictive" ones, namely, into messages occurring within a fictional space, operatic or otherwise. It was realized that music lacks a simple referent; it creates a new world of signifieds, which are well articulated and correlated. This fictive world of expression came into being with the creation of a hierarchical musical grammar—the grammar of tonal harmony—which served as a frame of reference for musical coherence. In the course of its formulation, musical language appropriated meanings by attaching them to well-defined musical components. Once these connections had become established, there emerged the possibility of their elaboration and manipulation in ways unique to music. On this basis, the development of instrumental music, liberated from the need for textual explication, was made possible. With the elaboration of the so called classical style, music's "possible world" was further enriched by a new resource: musical time became entangled with narrative time, to create a unique temporal experience.

How does such a symbolic system, which defies translation of any kind, act cognitively? It is instructive to note that while musical language was coming into being ways of thinking had developed which could deal with this question philosophically. Thus it happened that music, among other "natural languages," became the testing ground for plausible hypotheses concerning the working of the mind. The fourth chapter discusses the major outlines of this development, from Descartes to Condillac and Vico.

A group of British thinkers of the eighteenth century looms large in our book, because they perceived these phenomena in all their complexity. Moreover, they developed methods to address those components of consciousness, which are relevant to the processing of art-symbols in a manner enabling further developments along similar lines. By combining the Aristotelian conception of art as a kind of "making," with the Platonic conception of coherence as a primary aesthetic criterion, the British thinkers, we argue in the fifth chapter, created a new space for discussion, where the "qualities" of consciousness and the "modalities" of the artistic media (to use Cassirer's terminology) could be examined in their interrelationship. According to this group of British men of letters, the reciprocal relationship between "qualities" and "modalities," though dependent on their inherent dictates, initiate a continuous process of metaphorical exchange. The full comprehension of this process, they argued, must take into account cultural conditions and the changes they undergo. The medium does not forever remain bound to its initial message, nor is a given message forever limited to one medium only, as Lessing had argued.

If we are in the position to move beyond the thoughts of individual thinkers to the intellectual coherence of the British group as a whole it is because from our vantage point we observe how their ideas have been transformed, nowadays, into a unified conceptual system, which highlights the contribution of the arts and in particular music to theories of knowledge. Like Gadamer, we see no validity in isolating ourselves from our own time. Quite the contrary, it enables us to become fully aware of the basic premises of our consciousness. It is by means of this consciousness, we learn from Gadamer, that we are in a position to fully appreciate the significance of earlier historical moments that fostered the consciousness with which we are endowed.

However, as is well known, there are many ways of being part of a given historical moment, depending on different cultural codes, and on formulated as well as covert desiderata. The sixth chapter shows how the French and the Germans dealt differently with the questions that preoccupied the British, each nation in accordance with its own past tradition and tendencies. Rather than deny the value of their contributions to the analysis of the problems which emerged, it is perhaps their contributions, especially that of the Germans, that explain why the British contribution has been hidden for so long.

The concluding chapter repeats the main thesis adumbrated in earlier chapters, according to which the discussions dealt with in the book were necessary for the emergence of abstract art on the one hand, and of basic hypotheses concerning the mind, on the other. Thus, a basic problem which was clarified in the British deliberations on art concerned the relationship between process and coherence as related to construction or representation, which is influenced by that which is being processed, that is, by concrete sensual elements. This involves the question of modes of symbolization, which carry along their own systems of meaning and decoding, leaving a basic theoretical question pertaining to the relationship between perception and cognition still unresolved.

So much for *Tuning the Mind*. In *The Arts in Mind*, which we consider as a companion to this volume, we introduce a selection from the writings of the British thinkers, from Shaftesbury to Smith, in a manner easier for the modern reader to understand. The texts are accompanied by our interpretative annotations. Until now the British thinkers have not gained the honor they fully deserve, and their theories have only been partially discussed from the point of view of their inner coherence. Hopefully, our notes will enable the modern reader to follow the line of argumentation of these writers as well as to examine the rigor of our own arguments in the first part of the book.

* * *

Our whole-hearted thanks go to all those who have made this book possible, on all its levels. Some scholars mentioned in our book contributed to specific points, others to more general ones, and all together they created the arena of our discussion. As is inevitable in constituting intellectual history, the reader will not fail to notice that alongside works by modern scholars, we have also consulted canonical works of the past, which are still fresh. Some serve as a basis for both our own discussion and those of others, and some, which have not been sufficiently appreciated to this very day, are treated as if they were recent contributions. Our reference to Bukofzer's almost forgotten argument concerning musical meaning is a case in point. Gombrich, a more prominent representative of the Warburg Institute, is also referred to in this book, both for his contribution to the understanding of the activities involved in the construction of symbolic systems as well as for his research concerning their necessary conditions. We are no less indebted to the classical works of Hagstrum and Hollander in literature, Panofsky and Alpers in art history, E.T. Cone and C. Palisca in musicology, Max Black and Umberto Eco and many others in philosophy, semiotics, history, and in the various histories of the arts; they are all included in our bibliography. Above all we are indebted to Cassirer's work, whose presence in this book is keenly felt.

We dedicate the book to Carl Dahlhaus, who was a rare musicologist. His thinking in music was comprehensive, including many worlds of discourse, to whose interaction with music he was always sensitive. Dahlhaus insisted that one must be well versed in the musical material itself, but that it could not be self-explanatory. To our delight we had his blessing for this project and enjoyed the years through which he accompanied us, offering encouragement that we succeed in solving a problem, which he considered central.

1

Moving the Passions of the Mind: The Ascendance of a Query

This book promises to deal with a crucial moment in the deliberations on art in which interest in cognitive processes became prominent. That art involves the senses and the mind was always part and parcel of the assumptions that accompanied artistic activity, but art theory touched these issues mainly from a practical point of view, or in broad and universal terms. Even a complicated phenomenon like that of illusion did not stimulate an analysis of its necessary conditions. What then were the issues and problems that underlay the process of dispelling this "innocence"?

A review of the issues that gave rise to a new inquisitive orientation is necessary for the understanding of the deliberations on the arts that prefigured what may be called the cognitive turn. Such a review will also highlight the role of music in the elucidation of key factors in the working of the mind. The following pages will examine some of these issues and factors, and the theoretical reframing of the problems inhering in the relations between the senses and the mind which they instigated. In the course of this process, the relationship between Art and Nature underwent a fundamental change, as did the relationship among the arts, giving rise to new theories. It will become evident, nonetheless, that the new theories attempted to salvage whatever possible from old habits of thoughts—previously taken for granted—as they formulated novel frames of reference that were able to contain new concerns and artistic desiderata. Paradigmatic shifts, apparently, are the end products of prolonged processes of adjustments, inventions, and readjustments affecting each other in intricate ways. In this particular case, it involved no less than a reconfiguration of the seven liberal arts, which amounted to a renewed emphasis on those concerning language (the *trivium*), at the expense of the mathematical ones (the *quadirivium*). More particularly, it resurrected rhetoric as a mode of thought, highlighting the affinity between the future sister arts—painting, poetry, music.

Between Nature and Art: Relations Redefined

Among the many issues that invited clarification was the relationship between nature and the arts. Renaissance attempts to resuscitate the classical world involved the reinvention of old techniques, accompanied by the artistic conceptions that lent them support. Merging these ancient legacies with medieval ones required, however, new conceptual vessels. These inherited and reinvented legacies pertained mainly to poetry, painting, and music, to their relationship to nature, and to the relations among them. Foremost among what was inherited from antiquity, following Horace, were the premises that underlined the so-called *ut pictura poesis* ("as is painting, so is poetry") tradition[1] that aimed to emphasize the similarity between painting and poetry, stressing their common mimetic nature. Poets, accordingly, conjure up what is described through words, while painters depict *the same* through forms and colors. Poetry, however, throughout the Middle Ages, was part of the established educational curriculum, an art with a theory of its own, whereas painting was considered a mere craft. The *ut pictura poesis* dictum, and some related notions, constituted the attempt during the Renaissance to confer on painting the status held by poetry. Since the object described by painting is directly presented before the eyes of the beholder, it was believed by some that painting could even claim superiority over poetry, for its special ability to achieve the goal of re-presentation more directly (Summers 1987: 32-41; 125-7; 137-143). This was related, of course, to the classical origin of the dictum, which, as Gombrich indicated, imposed on painting the ideal achieved by Homeric poetry, that is, the rendering of historical events and mythological adventures as though taking place before the eyes of the spectator.[2] The *ut pictura* tradition thus imparted a double message: On the one hand, it set up as its goal the presentation of reality as *experienced*, granting *enargeia* (ἐνάργεια)—pictorial vividness—central place in art theory. On the other hand, by adopting an artistic model rather than reality itself, it implied the recognition of the powers of art to create artificial worlds that we experience as real. Imitation in that sense became likewise the goal in the Renaissance, leading to new visual and literary languages. Artistic norms were determined more and more by perceptual constraints, taking into account the ability of a beholder and a reader to conjure up "reality" out of pigments and words, emphasizing the ability of art to induce sensual and psychological imagery experienced as real.

Such a change in artistic goals is inconceivable, however, without societal transformations that give rise to new cultural desiderata. Indeed, the socioeconomic changes that accompanied the waning of the Middle Ages and beyond, were linked, as Katz argued, with the emergence of the new understanding of art. Briefly: the urbane commercial economy, which began to take shape in the late Middle Ages, led to the political and cultural emancipation of the middle class, and ultimately to its intellectual predominance. The allegiance to region and locality that accompanied this process clashed with the universalistic striving of the church and its attempt to maintain a uniformity of culture. It was the insistence on differentiation—of collectivity from collectivity, of individual from collectivity and individual from individual—that spans the period from the late Middle Ages to the Baroque. These changed conditions gave rise to the ethic of free competition granting recognition to those who "made it." In art, individual achievements were linked with the break down of the guilds, emphasizing the *creative* and *expressive* power of the individual artist, at the expense of shared artistic technique.[3] Artistic desiderata thus appeared more transient, diminishing the power of entrenched ideals and established norms. Yet what it is that art tries to create and express, and how it proceeds to attain these goals, became hence issues that gained in saliency.

The Musicalization of Poetry and Painting

All of the arts partook in the above developments, but unlike poetry and painting, music, despite Aristotle's attempts, was not generally considered among the imitative arts. The Platonic tradition, which attacked the imitative tendencies in art, had apparently the upper hand as far as music was concerned.[4] Considering its mathematical and physical components, music was related to world harmony as a static system, ignoring the fact that music is experienced as unfolding and developing in time. The basic Pythagorean conception, however, remained entrenched during most of the Renaissance, despite the Aristoxenian challenge, because Christian thought reinforced a harmonistic conception of the world, finding it compatible with its needs.[5]

Infused with Christianity, an important change did take place in this tradition, with repercussions not only for music, but for poetry and painting as well. Already in Plato's cosmology, harmonic relations unified all the parts that constitute the world, small and large, physical and spiritual. The church fathers adopted this metaphysical approach, but in line with the Scriptures. The combination of the cosmic order and the *anima mundi* gave rise to the Christian ideas of love and *caritas*, rendering the transition from external order to internal spiritual harmony natural. Adding "soul" to "harmony" infused otherworldliness to the worldly, unifying the act of creation. This tendency, however, found two distinct expressions, as Spitzer (1963) explicated—in the hymnology of St. Ambros and the writings of St. Augustine. The former tended to emphasize the spatial aspects of world harmony while the latter its temporal aspects; in both cases the sensuous element, so central in the pagan world, lost some of its standing. Thus, the spatial combined with the spiritual found vent in metaphorical language, for Ambros, while the temporal combined with the spiritual, emphasized, the dynamics of a narrative, for Augustine. If Ambros stressed simultaneity and an ever-present revelation, immediate and available at all times, Augustine emphasized change and process culminating in redemption. The latter, as a consequence, became more closely related to the "perfectibility of man" than the former, creating a greater awareness in man of his own actions (Spitzer 1963; Passmore 1972).[6]

Indeed "harmony" and its derivatives penetrated language and all of its by-products, so that they labored within an atmosphere thus "tuned," as we learn from Spitzer. This went, however, beyond Pythagorean notions. As far as the arts are concerned, the musicalization of painting is of special interest, for it enabled the eventual unification of the fine arts. While poetry and painting were related through *ut pictura poesis* practices—mainly through the genres of historical painting and exphratic and descriptive poetry—poetry and music were correlated through text-music relationships.

These were symbolized by the various practices of the muses, of which four had particular responsibility for genres of vocal music.[7] The bridge between painting and music could thus be erected either by making music into a more imitative art, or by leading the other arts to acquire some of its abstract harmonic conditions. Historically, music became more imitative, but only after the other arts, especially painting, underwent a certain kind of musicalization.[8]

The process started already in the Middle ages, when painters were no longer merely interested in the accurate rendering of sensual objects and tried, instead, to concretize that which is *in*visible, conceiving the visual image as an "introduction of the unseen—the supernatural—into the material" (Hagstrum 1958: 48-51). New interest in color and light—especially in Byzantine art—replaced the classic procedures of foreshortening and modeling in light and shade, that were developed in antiquity, to produce *trompe l'oeil* (Gombrich 1977: 108). The concentration on color continued during the Renaissance, in spite of the avowed aesthetics that hailed the classical ideals, which went hand in hand with the shift of emphasis from the beauty of form to the beauty of light. In this process "atmospheric painting" was born, especially in Venice, concentrating on inner reality rather than on natural form. In characterizing the fifteenth century the school of the Venetian Giorgione, Walter Pater, toward the end of the nineteenth century, made this hidden trend salient and saw fit to assess it as "meeting the condition of music."[9]

This trend found expression also in the "mainstream" painting of the Renaissance, as attested by the new language theorists employed in relation to its main achievements. Vasari (1568), for one, described the changes that took place in the visual arts from the early to the high Renaissance as a transition from *techné* to expression. Expression, he maintained, was *the* accomplishment of this third age (*etá*), which added "grace" and "air" to beauty, overcoming the "dryness" of the former periods. He might have learned from Castilgone (1527) that "Grace" stands in contrast to "affectation," and insisted that it should strike us through the "softness" or the "sweetness" *inherent* in the work.[10]

The new class of predicates that was attributed to works of art became gradually part of the discussion of the aesthetic experience, comprising one of its underlying dichotomies, that between "affect"—the quality of the experience—and "expression"—its manifestation in the work of art. The first, concerning the "subjective" aspects of art, would later be dealt with in psychological terms, whereas the second, concerning its more objective aspects, pertained to physiognomic accounts.[11] Renaissance art, as part of its iconographic tendencies, already developed physiognomies as a symbolic language; what had begun fortuitously, crystallized in the course of the Renaissance into recognized general moods and strains of particular sentiments that became associated with particular shapes and images. The laws of decorum—related to genres and subgenres—guaranteed intersections of ranges of meaning, excluding ambiguities and vagueness as far as deciphering was concerned, as Gombrich explained.[12] A young woman's head slightly bent and lit, eyes downcast and lips closed on the verge of a smile, one hand spread out as for blessing, the other firmly holding a reclining baby (in a painting like Leonardo's *Modonna of the Rocks*), this, even without iconic halos, was immediately correlated with the expression of transcendental, religious bliss.[13] Immediacy of this kind rested not only on conventions and habits, but also on certain psychological factors that guaranteed the inseparability of interpretation and observation. Later generations would recognize the combination of these factors as the very definition of this particular style.[14]

Unpredicated Signs in Search of Meaning

Music entered into more theoretical relations with her sister arts precisely when the interest in "expression" increased. Expression was from the start attributed to music, a fact that even Plato could not overlook. For Aristotle, as well as for a *pseudo*-Aristotle, this made music an imitative art, assisting poetry in modes of characterization, [15] while for Plato it gave rise to a rigid theory of ethos. According to Plato, certain musical configurations, related to mode, rhythm, instrument, corresponded with certain designated dispositions, of which Plato allowed (in his ideal state), only those encouraging soberness, courage, and liberality. The Aristotelian legacy was no less significant for future aesthetic approaches to music than was Platonism, looming large, however, only in the seventeenth and eighteenth centuries. In the meantime, what occupied musicians in the Middle Ages and the greater part of the Renaissance was neither the first version of expressionism nor the second, but rather the attempt to *rid* music of some of its extramusical connections. Making music an autonomous art involved conscious choices concerning the formation of its basic building blocks. This process entailed the isolation and organization of musical parameters, the crystallization of modal frameworks, notational clarifications and, above all, conscious attempts to gain control over the flow of musical time (Katz 1991). By the end of the Middle Ages, music was still wedded to text outwardly but it no longer needed its support.

Composers in the Renaissance continued the process; they elaborated the polyphonic techniques to the highest degree of perfection. The vertical as well as the horizontal dimensions of music were increasingly controlled; so was the construction of the work as a whole. But only when the final touches of contrapuntal devices were being accomplished did attempts to invest music's *unpredicated* signs with meaning gain saliency. Text and music were now beginning to be creatively intertwined, not merely carefully juxtaposed. Moreover, musical settings gradually ceased to be "a vehicle for the poem," and became, instead, "an independent creation that usurped the poem's message" (Palisca 1985: 371), as illustrated in Cipriano de Rore's madrigal quoted in example 3c below.

By the end of the sixteenth century, the new relationship between music and text became an explicit desideratum. Instead of music embellishing the text, words were now enlisted to pinpoint music's own signification. Despite the overinsistence on the importance of the text by members of the Florentine *camerata*, it was the recognition of the unique "message" of music which accompanied the development of opera, the cantata, and accompanied monody. Paradoxically, it was the process of realizing the ideas of the Florentines that brought music's own signification to the fore. Vincenzo Galilei (c. 1533-1591), one of their leading members, seems to have been aware of the nature of the new orientation—why else would he replace the "imitation of words" by the imitation of the *concetti* (conceits) of the mind? Lorenzo Giacomini (in 1576), Francesco Patrizi (in 1571), and other musical humanists spoke in similar terms. Nonetheless, the process of acquiring meaning in music achieved coherence only toward the end of the seventeenth century, once music had evolved an independent coinage for expressing the passions, that was embedded within a new tonal and rhetorical language.

Thus, while the visual arts had already appropriated rich semantics, the process of acquiring meaning in music was still in its infancy. Not that music was considered devoid of significance. On the contrary, due to its affiliation with Pythagorean theories, music and the powers ascribed to it were still closely related to the very structure of the universe in ways unequaled by the other arts. However, when Aristotelianism was reintroduced into musical thought, it encountered Platonic ideas that had already been affected by certain empirical touches. Already with Petrarch, transcendental expression was to be sought in Man, indicates Spitzer, and not in the dwellers on high (Spitzer 1963: 61-2). With Marsilio Ficino (1433-1499), a new attempt was made to understand the way in which the emotional reaction of the individual was linked to the "harmonious structures." Passions were awakened by musical means, which were still indebted to transcendental forms, while their actual configurations were increasingly determined by that which successfully "operates" on the hearer (Walker 1958; Katz 1986, Katz and Dahlhaus 1987 i: 75-93; Tomlinson 1993).

Ficino's speculations remained on the level characteristic of the Pythagorean tradition; they still accounted for the special ontological standing of music, rather than for its specific expressive contents. Yet experiments of the kind he made regarding the powers of music have gradually come to dominate the musical scene, overpowering its preestablished patterns. Whether magical or otherwise, such theories and experiments contributed to the crafting of effective musical means, resulting in the invasion of speculative thinking by practical considerations, themselves aspiring to a philosophical standing. This amounted to an increasing demythologization of music's ontological supremacy, which went hand in hand with the new attempts to invest music with specific meanings.

The Interpenetration of Rhetoric and Poetics

All of the interrelated changes discussed above reveal a growing awareness of the communicative potency of the arts. Yet this awareness was largely related to normative and practical aspects of art in the making, and not to their theoretical implications. Typical of this situation is the fact that the arts, which in spirit became closer to each other, were still located at their previous moorings. In the waning of the Middle Ages, painting, as already mentioned, was considered largely as a craft, and was thus held less than an art; music, by contrast, enjoyed a privileged place, on behalf of its Pythagorean standing in the *quadrivium,* thereby considered more than an art. Among the three arts, poetry was the only one to have theoretically benefited from its affiliation with logic, grammar, and especially rhetoric—the branches of the *Trivium.* In addition to that, it could directly profit from the recently translated *Poetics* by Aristotle, which reinforced its anchorage in a philosophical framework emphasizing poetry's power to create meaning and explicating the ways it goes about it. Poetry, hence, became the model dominating the entire scene, guiding painting as well as music in their attempts to extricate themselves from their confined surroundings. The turn to the very environment of poetry was thus achieved not only by actual alignments with literary sources but more importantly, by means of an associations with the branches that lend poetry its theoretical standing—classical rhetoric and poetics.

Poetics, since Aristotle, highlighted ways and means of imitation, introducing classifications among genres, and submitting them all to questions concerning verisimilitude and their purifying effect. Rhetoric, on the other hand, dealt with ways and means meant to affect the consumer in a direction desired. Each determined a different set of categories for criticism, directing the artist towards diverse goals (Abrams 1953: 1-29). Nevertheless, the two traditions were often intermingled. It was through the eyes of Horace, whose *ars poetica* was primarily dominated by the rhetorical outlook, that Renaissance scholars viewed the Aristotelian theory of imitation. Interested in Man and his moral and intellectual improvement, the Renaissance clung to the *prodesse delectare* (instruction and delight) dictum at the expense of issues related to "imitation," that is, focusing on artistic import rather than on the ways via which it is achieved (Weinberg 1961 i).

Rhetoric, however, had its own prescriptive tradition, applicable to discourse. In the Greek tradition, particularly in its culmination in the writings of Aristotle, rhetoric was conceived as a counterpart of dialectics, operating in the realm of non-demonstrative arguments. As an effective tool, adding validity and substantiality to forensic, political and other kinds of speeches, rhetoric was regarded as holding some truth-value. Not even Aristotle disqualified the speaker who exploited the emotions of the audience in order to achieve his purpose. Yet it is only with the Romans, primarily with Cicero, that rhetoric first came into prominence. Cicero was the first to establish the main parts of the art of discourse and to enumerate figures of speech and of thought as factors in causing pleasure and holding the attention and tension of the audience. Following Cicero, Roman rhetorical writings reveal an awareness of the ways of building and fulfilling expectations, taking into account the dictates of taste and propriety.

As a distinct body of knowledge, rhetoric became a cornerstone of the basic curriculum throughout the entire Roman Empire, partaking in the formation of a stylized culture for which lucidity, orateness, and appropriate diction became central values. It is all the more surprising, as Murphy points out, that the ancient world, "so productive of *artes rhetoricae*, should have produced so very few prescriptive documents in the realm of imaginative literature" (Murphy 1974: 27). Indeed, attempts to deal with literature from a rhetorical point of view were quite rare in antiquity, since this branch of knowledge was confined to the grammarians.

The Italian theorists of the sixteenth century differed from their predecessors, ancient as well as medieval. As Weinberg put it, their theoretical undertakings were directed to the transformation of the *ars poetica* and other non-poetical rhetorical writings, into a "total poetics" (Weinberg 1961 i: 109). Based on the "fairly fluid organization of the material" in Horace's treatise and the Ciceronian sources, they attempted to construct a more solid and substantial set of rules for the creation of a greater variety of poetical and literary writings. The attitude was partly pragmatic; as such it paralleled contemporary manuals for painters and sculptors, though it was more imbued with theoretical considerations.

Painters could thus turn to rhetoric and poetics in order to gain insights into their own craft (Alpers 1961; Summers 1987). Alberti may serve as an example, for he argued in 1436 that painting, like linguistic expression, aims to persuade and to evoke feelings, calling the artist to plan his work and all of its details like a speaker (Spencer 1957). Pictures were now "telling," "presenting," and "creating" a feigned reality no less convincingly than poetic works, and via their new theoretical underpinning, painters could convince the skeptics of the advantages of their art. Indeed, many theoretical attempts of the period were directed towards that goal. Leonardo's *Paragone* may serve as an example of the attempt to prove even the supremacy of painting over poetry, by associating painting with "natural philosophy," that is, with science.[16] Later in the century, however, it was agreed that calculations of proportions and the like may assist painting, but did not provide the theoretical foundation of the art (Panofsky 1968: 73-9). The intellect of the artist, wrote Federico Zuccaro in his *L'idea de' pittori, scultori e architetti* (1607), "must not only be clear but also free, and his spirit unfettered, and not thus restrained in mechanical servitude of such [mathematical] rules" (Panofsky 1968: 78). Towards the end of the sixteenth century, when science began to contradict "commonsense experience," as Ackerman maintained, and art, at least in Italy, was no longer interested in meticulous observations of nature, it was gradually understood that art and science must go their own ways (Ackerman 1961: 65; Summers 1987: 322-327). Poetics, consequently, thus maintained its unchallenged authority; it seemed, at the time, capable to provide more satisfactory answers to the problems that were raised.

The Rhetorization of Music

Compared with its sister arts, music availed itself of rhetoric seriously only at a later date, at the beginning of the sixteenth century. Since musicians and music theorists went through the same curriculum as did their fellow artists, one cannot account for their relatively late interest in the subject on the basis of the spread of the humanistic tradition alone (Vickers 1988). It is, rather, the special state of music at that moment—the accomplishments achieved in the art of counterpoint and the initial attempts to invest music with meaning—that gave rise to a renewed interest in rhetoric. The ancient art was now enlisted not only to explain what had been achieved, but also to define new artistic goals, suggesting means for their attainment.

The transfer of a conceptual system from one artistic realm to another is complicated, but even spontaneous analogies, like fresh metaphors, help to "select, interpret and systematize" the "facts of art" in ways which bring to light hidden facets and qualities (Abrams 1953: 31). In this respect, music theorists of the sixteenth century may be regarded as having suggested new perspectives. Rematching music with rhetoric, however, did not require a radical break, for the two were never really separated. Rhetoric, to begin with, deals, at least in part, with the musical parameters of language. "It is by the raising, lowering, or inflection of the voice," writes Quintilian in his famous *Institutio*, "that the orator stirs the emotions of his hearers, and the measure... of voice and phrase differs according as we wish to rouse the indignation or the pity of the judge." Music, in turn, according to the rhetorical tradition, is vested with emotional power precisely through its natural relation to affected eloquence: "Even by the various musical instruments, which are incapable of reproducing speech, different emotions are roused," Quintilian tells his readers (Quintilian 1983 i. 10 , 22-7; quoted in Vickers 1988: 372).

What is it then that music theorists and music practitioners tried to reclaim through professing anew music's relations with rhetoric? Were they seeking to reemploy the links with effective eloquence in order to gain reassurance with regard to the "powers" of music? If so, the gain would have been achieved at a heavy price, that is, the loss of music's acquired independence. Quintilian ascribed emotional power to instrumental music only because of its association with the musical qualities of speech! Moreover, unlike logic, which imposes upon language its own structures (as far as truthful utterances are concerned), and grammar, which determines its forms (guaranteeing "correctness") the third branch of the *trivium* presupposes the existence of language and thus of logic and grammar as well, enlisting them all for the art of persuasion. Did squinting in the direction of rhetoric reveal a longing for the revival of the old relation at the expense of music's hard-earned independence?

Ironically, it was linguistic models, together with grammar, which, in fact, assisted music in its emancipation into autonomy. It was not for the sake of likening music to language that medieval theorists used linguistic terms and concepts in their discussion of music. Rather, it was for the sake of elucidating music's own structures that theorists applied terms like "phrase," for example, to sound phenomena other than language. Once the segments of music were named and conceived of in some hierarchical fashion, they could lend themselves to further organizations, operations, and manipulations. Rhetoric, thus, became more relevant to compositional techniques, although it is not clear to what degree and in what way it penetrated Renaissance compositional practices. At any rate, the relations between rhetoric and music, including the actual transfer of terms and techniques, were not intended to undermine the autonomy music had gained. The infusion of "linguistic" traits into music contributed, in fact, to its distinctness.

Up to the sixteenth century, the interest in rhetoric was primarily directed towards practical goals. From there on, musicians turned to its theoretical premises as well, in the search for an overall new conceptual framework for their art. Cicero's model of communication as the union of *cor* and *lingua* was the main concern of the humanist musicians, as Vickers (1988) pointed out. According to Cicero, *lingua* related to the explicit contents of the text, while *cor* stood for the implicit mood it created. Creating proper moods or dispositions was the role of rhetoric, Cicero argued; for Calvin and others in the sixteenth century, it was music that primarily played this role. According to Calvin, no one can "be edified by the things which one sees without knowing what they mean and to what end they tend." Yet the word of God, he proceeded, finds its way to the soul not only through the literal message but also through song: for we know by experience "that song has great force and vigor to move and inflame the hearts of men to invoke and praise God with a more vehement and ardent zeal" (Strunk 1981 iii: 156).

The idea that music creates dispositions was not entirely new, as we have already seen. Since the days of the Greeks, when the "well-tempered" state was "tuned" by the "harmonious proportions," music was regarded as the appropriate art for the tuning of the soul as well. But this idea that was elaborated upon in the course of the Christian millennium was largely stripped of its metaphysical aspirations by the seventeenth century. Deprived of its transcendental base, music might have lost its rational and moral standing and have turned into a mere sensual trifle. It was thanks to the rhetorical tradition that the secularized "tuning of the soul" did not lose its claim to reason. Like texts, it was believed, music too involves mental operations, yet while text relates to "knowing," its rhetorical or musical component finds its way to the mind through the act of "moving." "The word of God," writes Thomas East in 1592, "delighteth those which are spiritually minded; the art of music recreateth such as are not sensually affected; where zeal in the one and skill in the other do meet, the whole man is revived."[17] Justifying music on either aesthetic or moral grounds, musicians, theorists, and religious reformers could avail themselves of these or similar arguments. That the "affections of the mind" constitute the province of music, became, in the course of the sixteenth century, almost a commonplace, though it carried along old philosophical underpinning, employed in novel ways. Vincenzo Galilei, for example, who shared Plato's distrust of the senses, attacked modern counterpoint for numbing the affections of the mind, that is for its lack of respect for the text and its affective message. Ironically, the cornerstone of Platonic thought—the weaving of consonances into a harmonious whole—was regarded by him as no more than the "tickling of the ear." Concurrently, the Aristotelian concept of "imitation of the conceptions that are derived from the words" assumed an almost religious status (Strunk iii: 122-125). In the long run, even this concept of imitation will be demythologized and the intricate links between "knowing," "moving," and "imitation of the affections" will be analyzed and explicated in cognitive, psychological terms.

In the late Renaissance, however, the distinction between "knowing" and "moving" did not entail an adequate explanation of the nature of either. But unlike "knowing" which was thought of in abstract terms, "moving" was treated metaphorically as well as literally. Paradoxically, the fact that literal explications rested on physical terms, as exemplified by Ficino's theory, highlighted "moving" as a process, calling for speculations concerning its elements and nature. In any case, the concept of moving the affections, from classical times to the eighteenth century and beyond, presupposed sympathetic relationships between the moving object and that which it moves. It has been claimed, that magical practices in the Renaissance, including those involving music, rested to a large extent on such a notion of sympathy (Foucault 1972, Freedberg 1989, Tomlinson 1993). To these presuppositions, the notion of the willingness to be moved, on the part of the recipient, was sometimes added. Still, as we shall try to show, there is a crucial difference between physical and metaphoric accounts of "moving."

Rhetoric, in any case, supported the turn from physical to metaphorical "moving." While the sky was being "untuned," depriving mind and music of their transcendental harmonious relations, rhetoric created an alternative frame for dealing with musical import, secular yet dignified in character, conceiving of dispositions as distinct mental entities. It granted aura and atmosphere aesthetic recognition, encouraging the search for their correlative artistic representations.[18] Instead of the simple submission to text requirements, or primitive notions about music's power, musical means were developed to emphasize the independent expressive potential of sounds and their logical combinations, as will be shown in chapter 3. Though texts were still considered crucial for "knowing," the reason had significantly changed, since words were less and less considered as competing with music in their joint adventure. Rather, it became evident that via the designations that texts lend to musical contents they can help clinch the latter's meaning.

The designative aspect of words, we shall see, will concern the late eighteenth century writers. Their reflections will be expressed largely in connection with two basic approaches to the rhetorization of music, which developed during the Baroque, that is, the recitative and the aria. The recitative—molding melody and rhythm in accordance with the "inflections and accents that serve us in our grief, in our joy, and in similar states"[19]—dealt, in fact, with "delivery," that branch of rhetoric which itself exploits musical parameters. With the accomplishment of that which was eventually subsumed by the Baroque aria, however, musicians and theorists became more and more engaged with *dispositio*, with rhetoric's accrued notions concerning the unfolding of orations and soliloquies. Aspects such as the holding of attention and of tension, the building and fulfillment of expectations, and the creation of unified wholes became central issues. No less relevant for the aria and related genres were the rhetorical figures and tropes that served to name, classify, and even develop distinct musical contents and procedures. All of these participated in the endeavor to lend music a sense of affective meaning. Still, during the sixteenth century the association of music and rhetoric remained mostly on the level of attitudes and beliefs, it rarely considered the transfer of the actual contents of one art to the other via prescriptions of structure, style, figures, and genres.[20]

The Ascendance of a Query

Towards the end of the sixteenth century one can discern a new theoretical orientation in theories of art. Whereas previous theories took the relation between subject and object for granted, theories now had to grapple with the question of "how it was at all possible for the mind to form a notion that cannot simply be obtained from nature, yet must not originate in man alone" (Panofsky 1968: 82). The idea that the notion conceived by the artist is identical with that which the beholder effortlessly perceives through the work of art, that is,the "immediacy" so gloriously achieved during the Renaissance, now called attention to that which makes it possible.[21] Indeed, the insistence on the ability of the spectator to understand a combination of truth and fiction "without undue mental effort" became a subject of inquiry (Lee 1967: 39). The readoption of the *ut pictura poesis* tradition likewise evoked interest in the underpinnings of "communication," that is,interest in the vehicle transmitting artistic intentions to various sets of sensations and images. The borrowing of poetical principles such as "instruction and delight" reflects a similar concern with communication. Artistic representation thus turned also into a philosophical problem, focusing on what is re-presented and how, and by what means it succeeds to refer to its object. However, the fact that theorists treated painting in poetical terms, and quite often, both through certain theological notions, created certain misunderstandings, which were hard to absolve. Transferring indiscriminately concepts and ideas from one art to another revealed disregard for the difference in artistic procedures unique to each of the arts.

This was partially an outcome of a metatheory established when neo-Platonic ideas, mingled with certain Aristotelian notions regarding "common sense" gained momentum, and managed to bestow upon beauty "sublime" value.[22] Beauty was said to be revealed to the artist by an act of emanation and to be perceived in the artifact through innate intuition. Influenced by prevailing poetic thought, theories of painting—such as those of the neo-Platonist's Giovanni Paolo Lomazzo—tended to regard artistic ideas as transcendental a priori forms. Imperfectly reflected in the physical world, Nature, it was believed, could be realized in its ideal form only through works of art.[23] Since such forms were considered innate, eternal, and of universal standing, questions regarding their actual formation in the mind were limited to order and merit. As Platonic forms, or even as Aristotelian ideas charged with medieval scholasticism, their sensual content was of relatively negligible value. As such they seemed to have blocked serious attempts to come to grips with the individual arts or with relevant historical and sociological factors related to their formation, though some awareness of unique stylistic imports can be discerned at the time.[24] Poet and artist, it was implied, are both guided by the same basic ideas embodied in different materials. Accordingly, the differences among the arts touch only upon their appearances, while their essences are shared.[25]

Still, the insistence on "ideal forms" reveals an awareness of the role archetypes played in artistic understanding. Likewise, artistic ideas more akin to Aristotelian thought—embodied in specific sensory means and modes (for example, Michelangelo's concept of the *concetto*)[26]—opened up theoretical options in the course of the sixteenth century, together with other Aristotelian notions, related to judgment, pleasure, memory, and common sense. These, according to Summers, were systematically investigated, prefiguring, as such, modern aesthetic theory. Summers himself admits, however, that they did not yet involve an aesthetic-cognitive query into specific forms and experience of art. Even Zuccaro, who related *all* thought to mental pictures, and all ideas to *created*, rather than ideal forms, presupposed a pre-established conformity between Nature, Divine Spirit, and the intellective mind (Summers 1987: 281-308). His view of the intellect as a "*tabula rasa*, a spacious smooth canvas prepared by us painters to receive all those figures that will be painted upon it, but in itself retaining no form, or shadow of form," despite its modern look, was actually an attempt of conceive of artistic design in naturalistic terms, of a general and universalistic bent characterizing concurrent neo-Platonic theories.[27] By the eighteenth century, when perception is brought into aesthetic theory as a factor to reckon with, the legitimacy of simple comparison among the different arts, and generalizations thereof, will have been called into question. Yet the linking of perceptual processes to understanding of abstractions made it, paradoxically, still possible for comparisons among the arts to continue. In fact, substituting *musica* for *pictura* in the *ut pictura poesis* was a consequence of this linkage, highlighting both common and distinct processes that are involved in experiencing the various arts.

Expressive Music: In the Wake of the New Science

Theories explicating the relationships among the arts and their affiliation to mental processes did not depend solely on understanding perception; they had to link perception to cultural forms and norms. The growing interest in expression emphasized the interdependence between the two. Music was central to this development, since the process of acquiring meaning in music was a more conscious one and its actual achievements more conspicuous than in other artistic domains, as it will subsequently become clear. Yet understanding the nature of this process in music had to wait until the process itself was completed.

Early seventeenth-century attempts to deal with the semantics of music faithfully reflect the intellectual ferment of the period. Still some thoughts that had crystallized in the course of the sixteenth century and before, survived into the eighteenth century and beyond. Revolutionary as the seventeenth century may have been with regard to the world of knowledge, it retained, nonetheless, some stifling archaic and ambiguous modes of discourse. Nevertheless, several persistent problems could now be approached with a different frame of mind, assumed by the new science, with new orientation and tools allowing for a "more accurate separation of the true from the false," as Descartes put it.

The pressure of organizing the ever-increasing theoretical body of knowledge in music came to the fore as early as the second half of the sixteenth century, as evidenced in the work of Zarlino. Marin Mersenne (1588-1648) and Athanasius Kircher (1602-1680) continued the enterprise, but unlike Zarlino's, and despite their antiquarian tendency, especially as far as Kircher is concerned, their works were thoroughly influenced by the "new method" which called for a new kind of integration. Like physics, music was virtually cut off from the discourse of metaphysics, reserving the transcendental for the "prime cause." In the sky as elsewhere, Greek animism was dead.[28] Vague analogies, the applications of which were unlimited and whose inferences unconstrained, were no longer tenable as a base on which sound knowledge could be constructed.[29] In music, the process of disconnecting mythologies from concrete practical considerations culminated in an increasing disregard of that order that once served to link music with the universe. As a subject of learning, music no longer functioned as another "book of knowledge" from which such inferences could be drawn. It became an object inviting definitions of its own, relevant only to its potential and actuality.

What kinds of definition did music invite as an autonomous field?[30] What were the positive aspects of the new scientific outlook as far as music was concerned? Kepler's *Harmonia mundi*, though still related to the harmony of the spheres, provided one kind of answer. Kepler, as is well known, dealt with the universe in terms of matter moving in space. Like his renowned contemporaries, he resorted to common denominators anchored in measurable units, employed not only in the definition of the objects but also in the formulation of the regularity of their appearance. Music in his model could retain its former privileged status in the *quadirivium*, since it could be quantified just like the other physical phenomena. Consequently, the limits of the consonant series could be explained and justified by Kepler astronomically, while astronomy could be illustrated through musical facts.[31] In less outstanding ways, the new outlook gave rise to inquiries concerning the phenomenon of sound, in relation to both instruments and the sense of hearing that were eventually relegated to the field of acoustics. In fact, almost all those who were interested in music at the time, whether from a philosophical, scientific, or moral point of view, contributed to this branch of physics. Descartes, Mersenne, Kircher, Huygens, and John Wallis exemplify the many that plunged into this fascinating field, soon reaching important results (Palisca 1961; Cohen 1984).

But the mathematical mode was not the only way to deal with music "scientifically." What distinguished the New Science was, above all, its unique combination of observation and experimentation. Combining observation with experimentation also affected musical circles; it determined, most of all, the thinking and activity that accompanied the rise of opera. The theoretical and the experimental were there combined, to bring about an effective relationship between music and text. Accompanied monody, opera, and cantata were the products of that "problem solving," serving as a kind of laboratory for the "powers" of music (Katz 1986). As a consequence, the aesthetics of music underwent a crucial transformation, functioning in a double capacity—it controlled the artistic production while serving, at the same time, as its conceptual tool. A new paradigm was thus ushered in, determining artistic and theoretical activities for more than three hundred years.[32] Still, the theoretical clarification which accompanied the "revolutionary stage"—in which the discussions of the Florentines were the most central—

could not assist in the analysis of its own assumptions, nor could it be helpful with regard to its implications; that was left for subsequent generations.

In retrospect, none of the scientific approaches could be theoretically satisfactory. While the first provided method, it left the very phenomenon of music uninvestigated, "decomposing" music into units divorced from aesthetic considerations and ideational content.[33] The second approach, on the other hand, involved the "composing" of that which needed explaining, that is,the very making of the phenomenon which it claimed to comprehend. In fact, it only suggested directions, highlighting those aspects that awaited rigorous investigations without unduly elaborating upon their relevance.

Why was the treatment of expression in music so problematic? What prevented the formulation of an overall theory of expression? The answer, hinted above, sounds almost trivial; the phenomenon, that is,*shared* and *stable* expressive coinage in music, hardly existed. The members of the *camerata* were cognizant of this, advocating, therefore, the kind of music which they believed had *existed* in ancient Greece (Walker 1941-2). On the other hand, Monteverdi's famous observation that "anger" has no counterpart in the current musical language, reflects the awareness that even ancient models would not suffice (Strunk 1981 i: 53-5). He invented *Concitato* in order to remedy the lacuna in the vocabulary, which he successfully employed in both his madrigals and theatrical works. Shakespeare, on his part, was careful in his *Twelfth Night* to invite music to take part in the dramatic action only where the musical means for expressing certain passions related to given situations already existed. They related more to general psychological effects—such as the effect of "antique song," the "dying fall" of a strain, or the "silly sooth" of song that "dallies with the innocence of love"—than to particular idioms.[34] At any rate, being aware of the lack of a musical language of expressions does not necessarily entail the realization that it is an artificial product. For Galilei, in fact, as for Fludd and Kircher, expressive music was considered a part of nature, equal in standing to the phenomena which science investigates. The belief that such expressive music once existed, enhanced the drive to unveil it, or at least, to *re*discover it. Art and science were unwittingly made to share precisely that on which they differ, causing no small misunderstandings in both theory and artistic practice.

Descartes—Between the Habitual and the Innate

It was, perhaps, Descartes who first recognized the difficulties that theories of musical expression are faced with. Rejecting the celestial "overtones," Descartes believed (at least in 1618) that the ability of music to arouse emotions could be investigated scientifically, since the phenomenon lends itself to experimentation and quantification. He searched for the universality of musical effects, and enlisted ad-hoc explanations for them, *isomorphic*: "A faster pace arouses faster emotions, such as joy, etc… the reason for this is that [the faster meters] occupy the senses more, since they are more things to be noticed in them"; *conditional*: "even animals can dance to rhythm if they are taught and trained"; or *perceptual*: "the steps are only a means to an end; they bridge the unevenness between the pitches of the consonant intervals; they do not themselves possess enough sweetness to satisfy the ear, but must be regarded in relations to their consonances…" (Descartes 1961: 15, 31-2.). However, contemplating the synthetic nature of musical effects led Descartes to realize that the musical message involved a complexity, both psychological and artistic, which no account of mere musical units could fully explain. Descartes had hoped to undertake a more thorough investigation of the subject in the future (ibid. 15, 52), but in a letter to Mersenne, written in 1630, he concluded that neither beauty nor pleasantness can have any definite measure. "What makes one man want to dance," he observed, "may make another want to cry." This, he argued, is because the same stimulus evokes different ideas in our memory; "those who have in the past enjoyed dancing to a certain tune feel a fresh wish to dance when they hear a similar one; on the other hand, if a man had never heard a galliard without some affliction befalling him, he would certainly grow sad when he heard it again."[35]

Shakespeare expressed similar ideas more poetically in the *Merchant of Venice,* when Portia speaks about the music heard when Bassanio undergoes the trial of electing the right casket: "Let music sound while he doth make his choice," she says, "Then if he lose he makes a swan-like end,/Fading in music. …/He may win,/And what is music then? Then music is/As are those dulcet sounds in break of day,/That creep into the dreaming bridegroom's ear/And summon him to marriage" (*The Merchant of Venice,* iii 2, 43-53). That same

Portia (in contrast with Lorenzo's essentialist ideas about the powers of music) would subsequently attribute the sweetness of a nightingale's music to the night which surrounds it, claiming that beauty is in the ear of the hearer and in the aura which context bestows on it, more than in its inherent qualities (ibid. v 1, 99-108).

There is no aesthetic value that is intrinsic to a work of art, imply these turn-of-the-century authors, and no specific expressive content constitutes an intrinsic part of music. It all seems to depend on a kind of conditioning,[36] on private connotations, irrespective of style, form, and genre. Still, Descartes's theoretical contribution to understanding expression did not stop with this negative conclusion. Having acknowledged psychological factors as part of man's epistemic system, the issue continued to occupy him. No longer interested in the aesthetic domain, he seemed to look, towards the end of his life, for a context in which shared forms of behavior could be observed, without having to resort to technical analyses of complicated artistic materials. His search resulted in the *Passions of the Soul* (*Traité des Passions de l'âme,* 1649), where he investigated the mechanism underlying emotional expressions. Though it ignored artistic problems, the treatise had a strong impact on artists and theorists of art and continued to exert its influence almost to the end of the eighteenth century. The new ideas the treatise advanced and the problematics it entailed led, however, to a misunderstanding of its arguments and to suggestions for their revision.

The separation between body and soul was a major premise throughout Descartes's work. Attributing to the body all that pertains to inanimate bodies (largely through their reduction into components such as "heat," "movement," and the like), he excluded from the realm of the soul everything except "thought" (Descartes 1985 §17: 335). In spite of some problems concerning the identification of "matter" with "substance," body and soul were conceived as two separate substances. Consequently, the individuality of Man seemed in danger, the more so since the intuition of unity, according to Descartes, was shrouded by the vagueness of subjectivity, unapproachable by the inquiring mind (Grene 1985, Hoffman 1986). Scholastic reasoning regarding unity he considered unsatisfactory; intuitions, he believed, should more readily be investigated via empirical observations. It is in this context, that Descartes turned to the passions of the soul, a phenomenon that provided a "case study" as well as a "laboratory" for the investigation of the body-soul relationship. What characterizes the passions, according to Descartes, is that in their case the soul is acted upon by the body, unlike other cases which exemplify the controlling power of the soul over the body.

Descartes argues that by dividing the emotions into passions of the soul and actions of the body, the Ancients must have been aware of the interaction that takes place between the two. However, contrary to what they believed, passion and action, argues Descartes, are one entity, since the passion of the soul is the action of the body, forming together a peculiar causal link between two different kinds of finite substances. How does passivity overcome the soul whose main characteristic is action? Descartes solves this problem by ascribing passivity to the understanding, and viewing understanding as an indistinguishable part of willing itself. Passivity, thus, constitutes but a change of emphasis in the regular behavior of the soul, not an essential transformation.[37] The rationality of man, even in his most irrational behavior, is consequently guaranteed. This, in fact, is the main theme of the entire treatise.

Relegating the passions to the soul, interpreting the activity they stir in terms of understanding, was not altogether a new idea, as far as the arts are concerned. As was indicated above, Vincenzo Galilei and his circle had already connected the "affects" of music to the "passions of the mind."[38] For that matter, all through the Platonic tradition, the mind was never divorced from affection, particularly with regard to music. The rhetorical heritage enhanced this tendency, and eventually replaced it, as we argued above. "Acting upon the *will*," on the other hand was a prominent Renaissance *topos*, based upon the new belief in the powers of the individual and his right to activate them (Katz 1986: 81-95). Conceiving man as individual as well as creator, it found expression in musical magic, both demonic and natural, but it was in natural magic that "will" and "passion" became causally related. Will "concocted" the magical formula while Passion bespoke its actual appeal (Walker 1958). The experiments and observations involved in these activities were eventually subsumed under the new aesthetic paradigm, the one related to the "divining" (in its double sense) of the powers of music.

The relevance of Descartes's theory to musical trials with expression may thus be stated more broadly: As in Renaissance theories of musical expression, particularly those in connection with the hermetic tradition, Descartes, too, understood passions in terms of a process. His picture of the process, however, was considerably more complicated than the one entertained by his predecessors. In Ficino's theory, for example, affectionate musical contents are transmitted to the soul via the ear and the animal spirits. The animal spirits, which consist, like music, of "airy" substance, are infused with the harmonic relations of the musical substance and communicate the latter to the soul through aerial motion (Walker 1958). The process is a determined one; Will only affects the choice of music. Despite his mechanistic outlook, Descartes presented a more complicated view, whereby the process is neither exclusively viewed in terms of "passion" nor in terms of "action," but as an ongoing interaction between the two.

Descartes's description seems to admit "interpretation." In fact, the whole process is unlikely to take place without it. Interpretation was also essential in Ficino's theory, in which the proportions delivered by the animal spirits, functioned as a code which the soul deciphered. Decoding, however, according to this theory, was considered involuntary, since it is "the hidden arithmetical exercise of a mind, *unconscious* that it is calculating," as Leibniz, for one, will later put it. For Descartes, too, the final stage of the process (that of recognizing the message), is a passive one. Yet he insisted on ascribing activity to the initial stage of the process, the one related to the interpretation of the emotional meaning of a given event. Since Descartes was aware of what we have termed "conditioning," it is no surprise to find that for Descartes the understanding of emotional content is distinct from the object or image to which it relates. Association, rather than an inherent bond, characterizes the relationship between an image and the emotion ascribed to it. The association is primarily determined by the initial encounter with the object, that is, by a contextual situation that brings forth a particular meaning, for which the soul, willingly or not, is responsible. This association is imprinted in the brain so that future experiences with the object will summon up the already established association (Descartes 1985, §42: 343-4). The fact that the association between image and emotion is habitual has important ethical consequences, which Descartes stressed, so is the association between stylistic factors and aesthetic evaluation, which he overlooked.

The soul, writes Descartes, may deliberately "will" to dissociate the image from its previously conditioned emotional connotations, in an attempt to link it to other connotations. Though "our passions...cannot be directly aroused or suppressed by the action of our will," says Descartes, they can be governed "through the representation of the things which are usually joined with the passions we wish to have, and oppose to the passions we wish to reject" (§45: 345). The wisdom of controlling passions lies in *in*directness, maintains Descartes, emphasizing the psychological possibilities as well as the moral necessity of such control.[39] Notice, that what may be controlled, according to Descartes, are not kinds of behavior but their actual occurrence.

The power of the soul is, nonetheless, diminished by entrenched habits. What starts as an arbitrary association ends in "possession":[40] "Once we have joined some bodily action with a certain thought, then one does not occur afterwards without the other occurring too" (§107: 365). This possession has far-reaching psychological implications, since it involves determined types of behavior. According to Descartes, it is the animal spirits that prepare the above association (of an event and its emotional content) as a code for their "own use" in future situations. This "code" determines not only the repetitions of a given association, but also the commands obeyed by the different parts of the body, whenever the association occurs. Different movements—whether of the blood or the muscles—thus become associated with different kinds of images and events. Unlike Ficino's, Descartes's animal spirits are pure mediators, establishing as well as activating the *link* between predetermined physiological behaviors and certain emotional states.[41] In the actual occurrence of the passion, they also partake in the duration and intensity of the experience, for they are capable of "maintaining and strengthening the passion...holding open or reopen the pores of the brain which direct the spirits to those same nerves" (§36: 342). Yet again, their activation is dependent on the will of the soul to so continue the emotional event.

Passionate behavior thus appears to consist of neutral patterns that become emotionally possessed. The awareness of emotional states is apparently demystified, in ways similar to those conceived later by William James and others, who used symptoms as indicators of states of being, rather than the other way around. Still, how movement (say, of blood) becomes identified with an emotional label (say, of love) remains a problem for Descartes. Acknowledging psychological circumstances, Descartes was nonetheless oblivious to social factors, which would later seem for a Wittgenstein crucial for emotional labeling. But for Descartes "shared forms of life" play no part in his account, only *private* associations, taking place in the very beginning of the emotional life, determine the identification of certain movements with certain emotional labels. When the soul is first joined to the body, explains Descartes, a bodily motion causes the soul to feel that particular emotion, which will later become identified as the emotion itself (§107: 365).

The epistemological premises underlying this argument have been, of course, questionable. Indeed, as Marjorie Grene rightly observed, the incoherence in Descartes's theory of the passions "might have defeated the Cartesian tradition in its infancy, had seventeenth-century metaphysics not seemed so alluring and seventeenth century 'philosophical psychology' such a tedious appendage to it" (Grene 1985: 42). Still, even if caught in the pitfall entailed in the mind-body dualism, Descartes's theory provided insights and suggestive ideas of great importance. A schematic summary of these can consist of the following assertions: (1) Passion is a *process* and as such entails four distinct components/stages—initial association, cognitive response, bodily reaction and a recognition of the soul. (2) The association is an act whereby external data are vested with "existential" meanings. (3) The cognitive response is unconscious, and yet responsible for both the habituation of associations as well as for the activation of the bodily reactions. (4) While types of bodily reaction are naturally given, this is not the case with their occurrences. (5) The ultimate recognition of a *type* of emotion is reserved for the soul, which grants passion a rational status. (6) The whole argument rests on the notion that external data, bodily reactions, and internal, emotional labels, are cognitively arranged into types, enabling the creation of a correlation between them. (7) Despite his inclination to categorize emotional behavior, Descartes is careful not to bestow on their external signs the epistemological status of natural kinds, which he confers on the passions. Though expressive signs are easily identified, they are not simple indicators of their expressed passions. Signs may in fact be "shared" by different, even contrasting passions, they may change in signification from person to person, and even be falsely employed (§112-136). In short, they are mediated by social and cultural factors, and relate to interpersonal, rather than to intrapersonal, communication.

We shall argue that all of these theoretical components revealed themselves in the making of the new music. Indeed, the more the "sky" was being "untuned," the greater was the gain for the actual "tuning" of the soul. It is ironic, therefore, that Descartes's theory of the passions, which contradicted his previous philosophical premises, especially those concerning innateness, should serve music and art theorists in the seventeenth and eighteenth century to fortify their Cartesian stance (Lee 1967: 29-30). Yet Descartes was not the only one at the time to suggest new theoretical possibilities to deal with these complicated matters. No less important for future discussions of music expressivity was the alternative approach advanced by Descartes's correspondent and friend, Marin Mersenne.[42]

Mersenne—Precursor of Isomorphic Theories

Unlike Descartes, who believed that it was within the power of reason to attain truth, Mersenne entertained a version of skepticism with regard to that possibility. Accepting the existence of a real world, he thought the ultimate knowledge of that reality to be inconceivable by human beings.[43] According to Mersenne, all we can conceive are laws related to our *experiences*, which enable their prediction. As systematic information about the world as experienced, the mathematical sciences, argued Mersenne, are not categorically different from those sciences that deal with Man himself. Though reminiscent of the Renaissance view, according to which no clear-cut separation between artistic and scientific activity existed, given the increased compartmentalization of knowledge in his own time, Mersenne's views should be considered, rather, as harbingers of the later reunification of science and the arts.

As far as music is concerned, there was no rift between Mersenne's experimental investigations of acoustical phenomena and his speculations regarding expressive music (Mace 1970: 9). The fact that aesthetic judgment is neither shared nor testable as a scientific judgement does not mean, according to Mersenne, that it is less reliable. One has only to define one's research objectives more rigorously, that is, to isolate the relevant basic variables and examine the interplay among them more carefully. Encouraged by his success in defining the nature of musical consonances (Palisca 1961), he considered it possible, at least theoretically, to bridge the gap between physical and perceptual data, even if the latter are of a more elusive nature. Musical consonances and music expressivity, accordingly, are not essentially different; in both cases, the need arises to account for unconscious processes that take place in the mind.[44] Unsatisfied with solutions such as the one provided by Ficino, he tried to find a new way to explain the transition from "sense to spirit."

Mersenne's solutions appear less convincing than his overall attitude. They may be viewed, nonetheless, as attempts to isolate the problems, while drawing the boundaries of their investigation. As we have already argued, both Descartes and Mersenne lived at a time in which the connection between affect and particular musical expressions had not yet been fully realized. Consequently, modes, genres, word-tone relationship etc.—the expressive components of Greek music—became the subject of Mersenne's investigations (Walker 1941-2). The fact that actual music *lacked* expressive coinage was undoubtedly a major factor in his attempt to find better ways to deal with musical expression and overcome, among other things, the traditional mode of discussing the subject, which failed to account for the relationship between certain musical patterns and perceptual mechanisms. Even practical recommendations, such as those suggested by Mei or Galilei, Mersenne believed, should be replaced, he believed, by more rigorous systematic considerations.

It is difficult to find the thread that links all of Mersenne's suggestions regarding the relationship of expression and affect. Chromatic steps and slow motion, Mersenne tells us, are essential to sad melodies; melodies should be constructed in accordance with the rules of rhetoric; the rhythmic nature of music should be examined carefully because of the emotional qualities it gives rise to, an so on.[45] In Mersenne's attempt to deal with affect one may discern, nonetheless, an idea concerning the relationship between musical patterns and psychological states. Dismissing as mere superstition the Pythagoreanism Baif and his Academy tried to imprint on French prosody, Mersenne ascribed the power of music to its undetermined movement. Affective musical movement, he emphasized, does not reflect "numbers," but rather feelings. "There is no more powerful means for arousing the passions of the hearers," writes Mersenne, "than using the same times and movements by which the same passions are produced in those who are moved by them."[46] The new investigation of music should thus consist of an inquiry into the nature of the passions, accompanied by an inquiry into the musical parameters that are analogous to their movement. Later we shall encounter an elaboration of this idea in Webb's overall theory of music expression.

The question still remains, however, what attracted Mersenne—a man of such vast intellectual interests—to this esoteric subject, that he returned to it, time and again, throughout his long career? Why was he so eager, despite Descartes's and Moreau's skepticism, to try and find a satisfactory way to deal with the issue? Mersenne, the cleric, it would seem, subjected science to pragmatic needs, and the latter to moral and religious considerations. The subject of affections, that is,the ways the soul is acted upon, was of central importance in this hierarchical structure. One should bear in mind that, the belief in a "natural" connection between words and objects was gradually waning at his times; language, was considered as no more than an arbitrary medium, the basic constituents of which are determined by convention. Only the *need* to communicate is natural; that need, however, may be realized in different ways.

It is in this context that Mersenne suggested that "music is the best of all possible languages." Due to the iconic nature of musical expression, music, he held, goes beyond the expressive power of natural languages, which have to rely on conventional (rather than iconic) symbols. It is not that he believed musical content to be more natural or universal, only that it has the power to "express with the greatest possible clarity and brevity the thoughts of the spirit and the desires of the will."[47] Here he introduces the argument with regard to the "isomorphic" nature of musical expression and affect: Intervals, like words and phonemes, says Mersenne, are arbitrarily chosen and culturally bound; so are certain rhythms and accents. "Although nature seems to give us the intervals of the diatonic scale, vis. major ad minor, and the half-tone in major, one could nevertheless use other intervals such as the sesquisixth and sesquiseventh etc... which may succeed very well... and the sequence of the intervals of the melody and of all music is artificial, for one cannot make use of it unless on has learned scientifically or by exercise or through practice" (Katz and Dahlhaus ii 1989: 101). These astonishingly modernist statements, that deny musical expression all innateness, attribute the universal basis of communication to certain propensities and dispositions, rather than to preestablished patterns. These propensities and dispositions are more salient in music than in other communicative systems, and hence can be more clearly and effectively acted upon. Music, argues Mersenne, "serves better for the moral life, and is more sutiable to morals than painting, for the latter is dead whilst music is alive and somehow transports the life, the soul, the spirit and the mental state of the singer... to the ears and souls of the listener" (ibid. 100).

Mersenne, no doubt, was an eclectic thinker, living on the brink of a new era. His ideas still mingle old and new premises, confusing at times a universal discursive-scientific language with that of an emotional language of communication, an error later repeated by Leibniz. Not wishing to forgo the dream of universal language, his quest was after that language which "without having learned the meaning "would make the same impression on all hearers" (Mace 1970: 15). In the final analysis, however, he looked for that communicative immediacy which constituted the base of the search of all of the arts since the Renaissance, laying the foundation for the investigation of music from that point of view. The direction of future explorations was thus set by Mersenne, even though he was not fully acknowledged as their harbinger, including the following presuppositions: (1) In order to understand the phenomenon of musical expressivity, one should consider musical movement in terms of the various musical parameters working together. (2) This movement should be conceived as taking place in the individual's mind, though it may be shared by all competent listeners. (3) Accounting for musical expressivity should be anchored, therefore, in a cognitive base relating to forms of recognition and activation. (4) The relation between affect and expression may pertain to two kinds of movements (musical vs. psychological) which are isomorphic. (5) One should, therefore, try to unveil the cognitive structures that make them "coincide," as Webb later put it.[48]

Whether Mersenne was or was not Rameau's forerunner as far as tonal theory is concerned (Gruber 1970), he certainly was a precursor of theories of expression that heralded the modern discussions of the relationship between music and language. Moreover, unlike Descartes and Bacon, who dismissed music because of its enigmatic character, Mersenne saw the advantage of music precisely in that character. Altogether, Mersenne's thought, to quote Mace, "points toward the coming revolutionary idea that the significant, valuable, and essential elements of man's experience are in the irrational areas of his being, not in reason" (Mace 1970: 3). Recognizing this, Mersenne went beyond classicist dogmatism and, like Milton, anticipated an era that would prefer the language of music to that of the spoken word, a language that could delineate and articulate the irrational parts of human thought and behavior.

Notes

1. There is a vast literature dealing with the various faces of this tradition, starting with Howard (1909) and continuing with Lee (1967) and many others. For a bibliography and survey of the thinkers who dealt with this legacy, from antiquity to modern time, see Markiewicz 1984.
2. "When classical sculptors and painters discovered the character of Greek narration," argues Gombrich, "they set up a chain reaction which transformed the methods of representing the human body... for the character of Greek narration as we know it from Homer is concerned not only with the 'what' but also with the 'how' of mythical events." The poet, argues Gombrich, is an eye-witness in the Homeric narrative and so is the painter in the "modern" Greek painting. "In the whole history of Western art we have this constant interaction between narrative intent and pictorial realism," summarizes Gombrich (1977: 107-113). Narration, as Auerbach's classic *Mimesis* (1953) explicated, leads to number of models of imitation, each structuring differently the imitated world. That Homer indeed became a model in this sense also to artists in the Renaissance becomes clear from an authority such as Petrarch who claimed Homer to be the "first painter of ancient memories" (Summers 1987: 141).
3. See Katz 1986: 62-75. It was Pietro Aaron who stressed the fact that composers, like poets, are born as such and cannot be taught the art of composition. See on this matter also Blackburn 1987: 274-8. As Summers explicated, the notion of the artist's creativity and individuality was associated in the Renaissance with his "physiognomy" reflected in his works of art—his *maniera*. This, in turn, relates to his *Spiritus* or *Pneuma*—ancient and medieval concepts of 'a kind of body taken by the soul from the planets', determining the individual's character and his destiny (see Summers 1987, chapter 6).
4. Gombrich considers Plato's ideas with regard to imitation in art as a reactionary opposition to the rise of "illusion" as a central precept of art (see fn. 2 above). There are good reasons to believe that the "Greek revolution," which, in fact, encompasses all artistic domains, did happen just before Plato's life time, argues Gombrich, and that his "outburst against the trickeries of painting was an outburst against 'modern art.'" Admiring the changeless forms of Egyptian art, which represent typical, conceptual "eternal" truth, Plato may have also longed for Egyptian musical practices that included only tunes of proper ethical standing (Gombrich 1977: 107-8).

   Plato's decrees concerning music in *his* ideal state, referring to the appropriate modes, rhythms and instruments, may be conceived in this light. So may be considered , for that matter, his Pythagorean conception of cosmic, harmonic structures.

Interestingly, in the *Timeus*, he ascribed his cosmological conception to ancient Egyptian wisdom. Almost bereaving music of sensual appearance (Lippman 1964) his conception became paradigmatic for the aesthetics of music heralded by the Christian church and beyond.

5. According to Pythagorean theory, the smaller the numbers that expressed the ratio of the frequencies of intervals, or of the length of the corresponding strings the more consonant the intervals. The distinction among sounds is thus given by nature, rather than judged by the perceiving ear. Unlike Pythagorean numerology, Aristoxenos, already in the fourth century B.C. presupposed that music is an activity of a collecting and building up of impressions in one's mind. "Since music is a successive production," he suggested, "its apprehension depends on two faculties, sense perception and memory." Musical cognition, according to Aristoxenos, implies the simultaneous discernment of a permanent and of a changeable element. The discrimination of particles, he observed, is part of "sense perception" while their "functions" are contemplated by the intellect (Aristoxenos, from the *Harmonic Elements*, in Strunk 1981 i: 27-30). Aristoxenos clearly turned the Pythagorean theory upside down, no wonder that the Church found the latter more compatible with its goals, since it emphasize God's given universal order rather than the vicissitudes of human consciousness.

6. Guillaume Dufay's *Nuper rosarum flores*—a motet composed for the inauguration of Santa Maria della Fiore's dome in 1436 (built by Brunelleschi) in Florence—may be viewed as embodying an integration, or rather transformation, of spatial, Ambrosian dimensions into temporal, Augustinian ones in a novel way. According to Warren's (1973) classical study, Dufy succeeded in projecting the harmonic proportions incorporated in the spatial structure of the cathedral (6:4:2:3) onto the structure of the motet—the temporal relations among its four major sections. From a musical point of view, harmonic proportions, hitherto embodied only in intervalic relations among pitches, could thus be felt on the temporal level, affecting new textural and melodic configurations. On the experiential level, the spatial/visual and temporal/audible incarnations of the same proportions in the motet could reinforce both Ambrosian synaesthic tendencies and Augustinian narrative ones. By the same token, it testifies to the predominance of neo-Platonic approach to art at the time, as expounded below.

7. The myth of the nine muses circumscribes a rich variety of text-music relationships: Erato was the muse of love poetry and lyre playing, while Euterpe was that of lyric poetry and flute playing. Polyhymnia, the muse of sacred music, carried a portative organ, whereas Calliope, the muse of epic poetry, was associated with tablet and a stylus. Interestingly, the muses, Mnemosyne's daughters, seem to have been "resurrected," in various embodiments, during the period discussed. For rich references relating to such embodiments see Davidson Reid 1993 ii: 680-5.

8. The musicalization of painting was also manifested through harmonic "numbers." Ficino considered the proportions of a thing preserved in the image as a reflection of the idea of the divine intellect as imparting to the image "something of the power of the spiritual essence which it embodies" (Gombrich 1972: 174). Leonardo, too, though Neo-Platonic ideas were of less avail for him, considered painting, like music, as a reflection of the harmony of the Universe (Gombrich 1986: 112). It may be that the preoccupation of painting with harmonic numbers came from architecture, for which harmonic proportions became central since the early Renaissance as attested by Wiitkover (1971) and exemplified in the Florentine example described in fn. 6 above.

9. In Venetian paintings, argues Pater, "all material details are so easily absorbed by that informing expression of passing light, and elevated, through their whole extent, to a new and delightful effect" (Pater 1980: 107).
10. See Barasch 1985: 215-224. Linguistic uses did support the view that the seemingly new expressivity was in fact a latent legacy inherited from the Middle Ages. Agathius, we are told by Hagsturm (1958: 49), a contemporary of Paulus uses "grace" (a theological quality of central importance in Christian thought) to describe the visual image of the "Prince of Angels" considering it as the root of its affective power.
11. According to Barasch, Pomponius Guaricus's *De sculptura* (1504) contains the earliest discussion of physiognomy in the context of art, but remains almost without followers during the sixteenth century. According to Summers (1987: 110-112; 143-6), however, cinquecento theorists tended to stress the naturalistic basis of physiognomy, the origin of recognized expressions in movements of the soul that were transformed into body movement. As we shall see, Descartes's *Passions de l'ame*(1649) went in similar direction, encouraging a new interest in the subject in the domain of art. For its implications see Ross 1984. Arnheim (1957) reintroduces the issue into modern aesthetic theory in connection with the centrality of images in the context of "visual thought."
12. See Gombrich's "Aims and limits of iconology" (Gombrich 1972). Relating to the literary criticism of D.E. Hirsch, Gombrich argues that genres are central to theory of symbolism in general and to iconology in particular. Gombrich stresses the "closure," on the interpretative level, of Renaissance works of art. The psychological attitude of Gombrich is significantly different from that of Panofsky (1983) who emphasized the conventional, arbitrary nature of Renaissance iconology.
13. That this was indeed the case can be learned from the Florentines' actual reaction to a similar work of art—Leonardo's *Virgin and Child with Saint Anne*. See Summers 1987: 126.
14. The concept of *facilitá*, which was in vogue during the high Renaissance, expresses a contemporaneous awareness of this stylistic achievement (Barasch 1985: 226). The term "immediacy" is borrowed from the British writers, especially Twining, for whom it played a central role (see pp. 223ff. below).
15. What the latter tried to explain is the cognitive base for the association between certain musical configuration and specific ethical (i.e., emotional) characteristics, highlighting "movement" as the factor bridging the two as we shall see below.
16. See Kemp 1989: 13-46. The tradition of comparisons—the so-called *paragone*—relates to an ancient literary genre which stemmed from the Greek passion for debate. The arts in general, claims Barasch, played a marginal role in these writings in ancient times, they began to play an important part only in the Renaissance, particularly painting and sculpture (Barasch 1985: 165 see also Summers 1987: 137-9).
17. Strunk 1981 iii: 162. Interestingly, religious reformers, like Calvin and East, promoted and exploited ideas already advanced by the Church Fathers, especially St. Augustine. Unwittingly, however, they contributed to the "secularization" of music, by stressing the *inherent,* rather than metaphysical, connection between text and music, religious as well as profane.
18. For a brilliant and thorough analysis of the cultural background related to the "untuning of the sky," as expressed in English poetry of the period, see Hollander 1961. See also Bush 1961; Grant 1985.
19. The phrase is taken from Peri's Forward to his *Euridice* (1601), explaining the method of singing, which he claims to have invented (Strunk 1981 iii: 14). The

invention, disregarding several nuances, was, however, an outcome of the new paradigm conceived by the members of the Camerta as a group and the "invisible college" which it embraced. See Palisca 1981; Katz 1986.

20. See Unger 1941; Butler 1980; Vickers 1988. An early example of the penetration of the "rhetoric of rhetoric" to music theory is Gioseffo Zarlino's *Le Istitutioni harmoniche* (1558). Though outwardly conceiving music in Platonic terms, Zarlino's theory is imbued with Horatian terminology.
21. Awareness of "immediacy" is revealed in various ways by contemporary artists and theorists. Pietro Aretino's theoretical approach is especially distinguished in this respect. See Barasch 1985: 251. The whole development is, of course, closely bound to the rise of a new audience which these writers—Florentines and well as Venetians—explicitly addressed (Barasch 1985: 241-262; Summers 1987: 125-141).
22. This development is extensively discussed by Weinberg 1961 (especially vol i: 250-296), where he discusses the defence of poetry by Neoplatonism.
23. See Panofsky 1968: 83-99, Barasch 1985: 232-5. The whole effort of Renaissance criticism to prove that painting, like poetry, was an imitation of ideal truths, was in the main, according to Lee (1967), anti-Platonistic, though the tools used by the contenders were, ironically, of Platonistic origin.
24. The most explicit expression of such awareness is to be found in Vasari's work. See Barasch 1985: 209-228.
25. Even deliberations regarding the priority of sight over the other senses was mostly related to its transcendental origin. On the whole, this attitude, despite eighteenth-century attempts to refute it, never left the stage of aesthetic thought and is to be found at the basis of nineteenth- as well as twentieth-century "correspondences among the arts," which, while ignoring the perceptual dimension, prefer to deal with some general shared concepts and ideas.
26. This is best expressed in his famous sonnet "*Non ha l'ottimo artista in se alcun concetto/Ch'un marmo solo in se non circoscriva*" [Not the best artist had in himself any concept that a single marble does not enclose in itself] in which the artist's formative idea is seen as already implanted in the matter, that is, bound to the medium in which it is to be embodied. It is no wonder that already Michelangelo's contemporaries interpreted his artistic conception in Aristotelian terms. See Panofsky 1968: 115-121. For other uses of *concetto* by Renaissance writers see Summers 1987; of interest are Paleotti's ideas, but even he, despite some insights concerning the power of artistic images, still does not link their specific sensory condition with their particular meanings and appeal (ibid. 146-150).
27. Summers 1987: 288; Summers's interpretation of Zuccaro's ideas follows both Panofsky (1968) and Barasch (1985) in emphasizing the author's deep Aristotelian bent, which he explicates in greater detail.
28. Grant 1985. In Fludd's musical theory, inspired by the "musical magic" developed by Ficino, Agrippa, and Giorgi, *anima mundi* is the "*praecipuum agens* establishing the harmony of the universe." Without its inspiration, argues Fludd, "there would be no consonances in the *harmonia mundana*; neither any tones nor any coherent tonality in the world." It is precisely this animism together with other "oratorical imagery" which raged Mersenne, who tried to purify the skies of all metaphysical entities and to free the soul—the human soul—of magical influences, demonic and otherwise (Amman 1967: 216-8; 226. See also Debus 1979).
29. On the changing status of analogies during the scientific revolution see Toulmin 1961, Foucault 1972.

30. Descartes's preface to his *Compendium* of 1618 provides such an autonomous definition—Aristotelian definition tinted with rhetorical hues, foregoing the still prevalent theological and metaphysical aspirations: "The basis of music is sound," writes Descartes; "its aim is to please and arouse various emotions in us." By the same token he distinguishes between music and acoustics, claiming that "quality of sound...is the domain of the physicist" (Descartes 1961: 11).
31. Several authors dealt with Kepler's musical-astronomical speculations, the most important of which are Walker (1978) and Koyré (1961). For a short introduction to his theory and an English translation of his main arguments see Katz and Dahlhaus 1987 i: 107-140.
32. Behaving almost like a "circle" of scientists, the Camerata members focused—as part of their varied activities—"on a problem (how to make language attached to music audible); they agreed on a methodical approach to its solution (adapting music to words or words to music: the direction was in dispute); and they agreed on the types of phenomena to be manipulated and observed (the power of music "to move the passions of the mind"). These are Kuhn's (1962) terms for describing the directives in a paradigm: identification of problem, methods and phenomena to be observed" (Katz 1986: 3).
33. It was Fludd (considered by the new scientists as an old-fashioned "hallucinator"), who obsessively stressed the irrelevance of theories advocated by Kepler and other scientists to questions related to the "power of music." His long debate with Kepler, which later also involved Mersenne, is perhaps one of the most interesting that took place at the rise of the scientific revolution. "Even though at the cost of consciousness of the quantitative side of nature and its laws, Fludd's 'hieroglyphic' figures," writes the physicist Wolfgang Pauli, "do try to preserve a *unity* of the inner experience of the 'observer' and the external processes of nature, and thus a *wholeness* in its contemplation—a wholeness formerly contained in the idea of the analogy between microcosm and macrocosm but apparently already lacking in Kepler and lost in the world view of classical natural science" (Pauli 1955; quoted in Amman 1967). For further explications of Fludd's theories see Heninger 1978.
34. See Hollander 1957: 74-77. The sensitivity to the limits of music (then current) to express the emotions is interestingly exemplified by the rejection of Jan Ban's "catalogue" (1637), which aimed to vest music with every shade of sentiment, discriminating even between motifs attached to "martial vigor" and those related to "indignation." Mersenne and Descartes considered his system as fussy, mechanical, and mediocre. See Pirro :1907 110, also see Rogerson 1945. Two generations later, when musical idioms had been established, such catalogues became more popular. See Dammann 1967.
35. *Descartes's Letters,* see Katz and Dahlhaus ii: 525. Interestingly, in the *Compendium* Descartes pointed to an *intrinsic* expressive musical characteristic, which later was applied precisely to distinguish, from an expressive point of view, between a pavan and a galliard: "As regards the various emotions which music can arouse by employing various meters, I will say that in general a slower pace arouses in us quieter feelings such as languor, sadness, fear, pride, etc. A faster pace arouses faster emotions, such as joy, etc. On the same basis one can state that duple meter 4/4 and all meters divisible by two, are of slower types than triple meters, or those which consist of three parts. The reason for this is that the latter occupy the senses more, since there are more things to be noticed in them." Given such a psychological reasoning, one may really wonder about Descartes's later position, expressed in his

letter to Mersenne. Perhaps the reason for his change of position is hinted in the last sentence of the quoted paragraph: "But a more thorough investigation of this question depends on a detailed study of the movements of the soul." It may be that Descartes was not satisfied with contemporary attempts to correlate the psychic and acoustic "movements" and therefore refused later to deal with the whole issue. As for subsequent application of metrical distinction in theories of expressivity in music see Allanbrook 1983.

36. This reiterates the conditional theory hinted at the *Compendium.* To be conditioned by music, thus, relates to its lack of decisive cognitive content.

37. This problem was first discussed by Descartes in his letter to Regius, 1641 (cited in Grene 1985).

38. Galilei's terminology, which appears in his *Dialogo*, is also used by Henry Peacham in his *Garden of Eloquence* (1593). For full references see Vicker 1984:12. Also see Strunk 1981 iii: 22-3 for Caccini's terminology.

39. See, for example, §46, and onward. It is the soul that determines the initial association between image and emotion, yet its power diminishes with the increase in habituation; § 42, 49.

40. "Possession," according to Goodman (1968: 50-52, 68-70), relates to a mode of symbolization in which an object, by means of an attached "label," exemplifies *only* those properties it both has (i.e., possesses) and refers to. This mode of symbolization is particularly relevant to artistic expression, argues Goodman, the only difference being that in art what is expressed is *metaphorically* exemplified.

    It is the "existential" meaning of the associated image or event that comprises, in Descartes's theory, the "label" identified with the psychological experience or behavior.

41. The animal spirits underwent a decisive change on the way from Ficino to Descartes—they lost their "spirituality" (and thus their quasi-metaphysical character), and were reduces into a bodily essence. According to Descartes, their peculiarity lies in the fact that they are "being extremely small" and "move very quickly, like the jets of flame that come from a torch" being thus a very efficient mediator between brain, nerves, muscles, blood and the pinal gland (§10, p. 332). As such, they will continue, until the end of the eighteenth century to play an important theoretical role, even if their empirical status will diminish to the bare minimum, as is the case with Daniel Webb (Arts in Mind: 260ff.).

42. The relationship between these two contemporary philosophers-scientists commenced in the Jesuit institution in La Fleche and continued all through their life. For the list of letters between the two concerning musical matters see the American Institute of Musicology edition of Descartes's *Compendium of Music* (Descartes 1961).

43. See Popkin 1960: 132; 1967; Mace 1970: 24-5. Mersenne's anti-solipsistic outlook may be exemplified in his treatment of music, where he unequivocally states that the phenomenon of sound exists irrespective of the existence of senses (*First Book on the Nature and Properties of Sound* in the *Harmonie Universelle* 1636). See Katz and Dahlhaus 1989 ii: 82-84.

44. It is amazing how close Mersenne's account of the process underlying the perception of consonances is to that expressed in Leibniz's famous phrase concerning the soul's "unconscious counting" while perceiving music (quoted above)."One can say" argues Mersenne, "that hearing is nothing but the counting of beats of the air whether the soul counts them without being conscious of it or whether it feels the numbers that touch it" (Katz and Dahlhaus 1989 ii: 90).

45. All of these points were raised by previous theoreticians. Zarlino, for example, pointed to that which characterizes sad melodies (Strunk 1981 ii: 67); Peacham, Bacon, (Butler 1980) Burmeister (Palisca 1972) and Descartes, among others, addressed the rhetorical components of melodies; Bacon also gave account of the perceptual and psychological components of rhythm. Indeed, Mersenne's novelty lies elsewhere, as we are trying to argue.
46. *Harmonie Universelle,* "De L'Art de Bien Chanter," Partie III, Prop. xxvi: 402-3; quoted in Mace 1970: 21.
47. Katz and Dahlhaus 1989 ii: 89. Taking into account the last points as well as fn. 41 above, one might argue that Leibniz was interested in Mersenne's ideas because of his interest in a "universal language;" See Couturat 1961.
48. Two more quotations from Mersenne may buttress our argument. Melodies, argues Mersenne, can represent "the diverse movements of the sea, the heavens, and of other things in this world, as long as one can keep the *ratios* in the musical intervals as are encountered in the motions of the soul, of the body and the elements and the heavens (Katz and Dahlhaus 1989 ii: 100). Subsequently he writes: "the nature of sad tunes...consists of several things: because the voice of the sad tunes represents languor and trembling; and the half tones and sharp tones represent weeping and sighing by reason of their small intervals which signify weakness; since the small intervals produced in ascending and descending resemble children, old people and convalescents from a long illness, who cannot walk quickly and make little way in a long time; for example when one goes up by one halftone in minor, one advances on one's road by only the twenty-fourth part of the preceding sound" (ibid., 103).

2

From Harmony of the Spheres to Music's Atmosphere: Trends in Seventeenth- and Eighteenth-Century Art

As we have seen, reason did not displace emotions with the new understanding of science; it was enlisted not only in the service of scientific truth but also in order to rationalize the a-rational. Moreover, there arose an awareness that expressions must be embedded in languages of their own, in systems of signs not yet in use for other denotative purposes. Our analysis, therefore, will trace the experiments with expressivity in music and the sister arts in order to extrapolate some essential processes and achievements that seemed to be a prerequisite for the later deliberations on art.

In the following chapter we deal with new trends in painting and poetry, which reveal the new sensibilities. Since eighteenth-century theories mainly deal with the relations between music and poetry, we saw fit to explicate seventeenth and eighteenth century poetical trends more than their pictorial counterparts, and to limit ourselves to the developments of literary genres and forms, more relevant to our British thinkers, rather than engage the entire European literary scene. The developments dealt with in this chapter are dominated by two seemingly contradictory tendencies that are mutually related. The first pertains to the openness of artistic forms and concepts, the second to the metaphoric interactions among them. Openness implies indetermined boundaries, which may obscure identities, including that of the medium. Interactive metaphors among the arts, on the other hand, demand clarity concerning the entities involved. Borders change in the course of the interaction, as do former identities. While such processes presuppose openness, inviting novel artistic procedures, they invariably induce new conceptual and perceptual challenges, increasing theoretical awareness. This, in a way, summarizes the narrative we are about to delineate.

From Closed to Open Forms

Though theoretical explorations of pictorial expressions were primarily motivated by the desire to equalize *pictura* and *poesis*, painting provided theory with more explicit expressive formulae than did the other arts. Different schools of painting agreed that expression must crystallize into shared recognizable patterns. It was increasingly emphasized, that "ideal nature" must be sought not only in nature but in works of art as well (Pevsner 1940; Panofsky 1968; Summers 1997). In Italy, painters continued to explore the whole gamut of human expressions: Caravaggio's genre paintings, like the "caricatures" of the Carracci brothers, "reformulated" an imperfect world through the dramatic highlighting of its latent expressivity.[1] In France, painters were engaged in applying predetermined patterns to actual expression; Charles LeBrun's was the epitome of this movement, which bordered on academic pedantry. Descartes's theory of the passions—especially his identification of certain behavioral patterns with specific emotions—supported this attitude, as was already mentioned (Ross 1984). And in the wake of the new ideas of the self, promulgated by Michel de Montaigne, painters like Poussin and Rembrandt were intensively preoccupied with their self-portraiture, attesting to their new understanding of the vicissitudes of emotional life and their belief in their power to express them.[2]

While this was taking place, seventeenth-century artists continued to seek universal essences in the atmosphere that surrounds their objects and in the feelings those objects evoke, submitting fresh observations to general laws. The genre of landscape, that developed in the sixteenth century and reached an apex with painters such as the French Claude Lorraine and the Dutch van Ruisdael in the seventeenth century, exemplifies this search. Substituting expression and atmosphere for proportion and harmony, painters were more ready to acknowledge the centrality of archetypal patterns in art. In fact, one might argue that the elaboration of certain artistic procedures such as perspective and chiaroscuro acquired as much awareness of shared cognitive structures, as did the development of techniques to convey affects and atmosphere. Yet, in order to communicate specific emotional states, the visual forms to which they related had themselves to turn into recognizable designators. Certain contours, tone-patterns, and compositional devices were deliberately chosen, since the expressive message seemed to depend upon them.

At the same time, the changing ideal in painting carried stylistic implications that went contrary to the search for recognizable designators. Emphasizing ephemeral essences such as aura and atmosphere, seventeenth-century painters tended to depreciate the separating, defining lines, replacing self-contained units with open forms as Wölfflin argued long ago. Light was used to dissolve form, not to construct it, and figures appear in the scenery only to re-merge with their environment. The visual representation of atmosphere and feelings so beautifully represented, as we have seen, by the Venetian school, acquired a set of stylistic procedures different from those which were used for rendering objects. It is this trend in the visual arts (as noted in chapter 1), that most clearly "aspired to the condition of music," that is, to the achievement of a "perfect identification of matter and form." In the seventeenth century, this trend wandered northward, finding new embodiments in the painting of La Tour, in France, Rembrandt and his school, in Amsterdam, and many others. According to Spitzer (1963), the set of beliefs concerning a well-tempered world in which body is tuned to soul, man to man and to his environment was closely related to the idea that a unique atmosphere, a certain *Stimmung* (mood) surrounded the participants.[3] In the seventeenth century, when culture at large was being "untuned," the emphasis on these ideals, albeit through metaphors, kept this tradition alive and intensified it. Among other things, it seems to underlie the selection of "ideal instants," of "exquisite pauses of time from the feverish, tumultuously colored world," rendering them in a way that conveys the "consummate extract of the quintessence of life."[4] A late blossom of the old *ut pictura poesis*, atmospheric painting encouraged the future replacement of *pictura* by *musica*.

Expression and atmosphere were not, however, the sole interest of seventeenth-century art; indeed, Italian painting lost its hegemony during this period. Dutch art strongly put forward an alternative set of artistic norms to compete with those of the South (Alpers 1983). Whereas the Italians were engaged in rendering significant human actions after the model of poetry, the Dutch painters were interested in the "prose of the world." They viewed symbolic representations as irrelevant to painting and considered ideal nature as a misleading standard. For the Dutch, who considered the senses—primarily sight—as the foundation of all knowledge, painting achieved an almost scientific status.

To be sure, already in the Renaissance, the Italians considered "seeing" as a means of obtaining knowledge, but the idea of knowing through seeing became less naive with the elaboration of the microscope and other optical devices, developed mainly by northern scientists. It became clear that a simple reflection of the world as it is, is impossible, and that "seeing" pertains to different scales and modes of observations, each of which entails a representation determined by the nature of the "instrument" used. Whereas the Italians considered the picture "as an object in the world, a framed window to which we bring our eyes," for the Dutchman the picture contained "the eye within the frame leaving the location of the observer undefined" (Alpers 1983: 45). Dutch painting, thus, called for a new kind of perception of art. In order to perceive the picture, one had to position oneself within it, following the movement of the eye that the picturing determined (See pictorial examples 2.1 and 2.2).

Poetical and narrative ideals were thus replaced by the intricacies of sight. Not only was the "*ut pictura musica*" discarded, but the musical qualities of painting seem to have been discarded as well. In that sense, Dutch painting developed in the opposite direction from that of the South. Dutch painting undermined the association of painting and poetry in yet another way: Though it identified perception with images presented to the mind in unmediated fashion, it highlighted interpretation as a central feature of perception. What seemed, on the surface, an immediate reflection of the world, involved, in fact, an active adjustment on the part of painters and viewers alike. A dynamic mode of observing pictures replaced a static way of perceiving them. This conception was later developed in the context of music and poetry, but the first intuition in this direction may have come from painting. The same may be said with regard to notions concerning the alliance between science and art: Dutch painting, like science, furthered knowledge by fabricating new modes of observation and perception. It may thus be viewed as yet another path to the eventual consideration of "discovery" as the

Pictorial Example 2.1
Raphael: St. Cecilia

Pictorial Example 2.2
Jan Vermeer: The Music Lesson

creation of systems, and "creation" as the discovery of how systems may be used—an understanding endorsed by writers in the second half of the eighteenth century.

Eco considers these tendencies as characteristic of Baroque art in general. "The various devices of perspective," he argues, "were just so many different concessions to the actual location of the observer in order to ensure that he looked at the figure in the only possible right way, that is, the way the author of the work had devised various visual devices to oblige the observer's attention to converge on" (Eco 1981: 50). On the other hand, he continues, "baroque form is dynamic; it tends to indeterminacy of effect (in its play of solid and void, light and dark with its curvature, its broken surfaces, its widely diversified angles of inclination); it conveys the idea of space being progressively dilated. Its search for kinetic excitement and illusory effect leads to a situation where the plastic mass in the Baroque work of art never allows a privileged, definitive, frontal view; rather, it induces the spectator to shift his position continuously in order to see the work in constantly new aspects, as if it were in a state of perpetual transformation." For the first time, he argues, "man opts out of the canon of authorized responses and finds that he is faced (both in art and in science) by a world in a fluid state which requires corresponding creativity on his part" (ibid. 52).

Whether or not this new "outlook" was shared by all, it seemed to contradict the belief in "ideal nature" that nurtured the search for expressive languages. Later theorists were aware of this seeming contradiction and tried to overcome it, though not, necessarily, in the context of painting. The intensity with which they sought the limits and constraints of representation was common to almost all schools and genres in seventeenth-century visual art. Some painters, like Poussin, were occupied with texts, and incorporated the interaction between media within the precinct of a single medium.[5] Others were engaged with vivifying emblems, employing *impresa* and other ready-made pictorial symbols. Some, like Velázquez, Gerard Dou, or Vermeer inserted the painter himself into the picture, toying with the dual purpose of presenting the re-presented.[6] Such a play was also highlighted by those who were led by their "mapping impulse" (Alpers 1983: 119-168), or by rendering pictorially architectural designs. At the same time, a painter as distinguished as Rembrandt settled for the materiality of his medium; "profoundly mistrusting the evidence of sight," he turned from the art of describing, Alpers argues, to that of evoking time passed. Altogether, the conditions and status of illusion were challenged; the relation between the fixed and the unfixed, the given and the made and the interaction between them, would, in fact, turn into a major issue. The poetry and the literature of the period already point in this direction. The following pages will deal with developments in literature, delineating the ways in which it came closer to music.

Redefining the Power of Language

Unlike painting, the masterpieces of which where shared by the European educated elite, and music which was viewed as a universal language, developments in literature did not so readily cross national boundaries. Our discussion of literature is hence limited to the prevailing trends in England, since it is these trends that eventually engendered the theories with which we are concerned. "Musicality" in this literature achieved in different and even contrasting ways, promoted the new interest in the correspondences between literature and music.

What we understand by musicality, as far as English literature of the seventeenth and eighteenth century is concerned, is different, though not altogether unrelated, to what is usually associated with the concept when employed in connection with romantic and symbolist poetry. The musicality we discuss is related more to the newly conceived powers of language rather than to its shortcomings as exposed in the later period. The clash in this period with the new scientific method and the philosophical climate that surrounded it, we maintain, called attention to the ability of language to activate mental faculties in various ways. In the differentiation between scientific and poetic language, which it gradually introduced, the power of metaphor and the rhythmic resources of poetry gained credence, while its descriptive capacity became suspect. Why all of this should bring poetry to the province of music will become increasingly clear as our narrative unfolds. As we shall see, the recapitulation, in the second half of the eighteenth century, of insights brought forward at the beginning of the seventeenth century, constitute a "theme with variations" of these issues. The final part of this chapter, which discusses the rising novel, takes us back to the beginning of the chapter, aiming to show how the novel, in both senses of the word, incorporated the conception of representation which were advanced by seventeenth-century Dutch painting.

The new tendencies anticipated certain conceptions that we have ascribed to theoretical developments in the eighteenth century. This anticipation rather than weaken, enhanced our argument, for there is an essential difference between the separation of science and art in the eighteenth century and the separation of the same in the preceding one. It is a difference between what Eco calls a "theoretical level of aesthetics," which tries to formulate definitions of philosophical positions, and a "practical level of poetics," which offers a programmatic guide for creation. It is through aesthetics alone, Eco tells us, "that the latent possibilities of a certain type of experience in every artistic product, independently of the operative criteria which presided over its moment of inception," are revealed (Eco 1981: 64). To put it somewhat differently, whereas the practical level is a stage in the natural development of art, the assessment of its theoretical significance requires hindsight; the theoretical level consummates, as it were, the historicity of momentary achievements, granting to some fundamental and lasting influence.

From Linguistic Norms to Linguistic Powers

Indeed, as far as poets and men of letters of the seventeenth century are concerned, linguistic norms were supposed to govern all discourses regardless of type. Though thinkers like Thomas Hobbes and Thomas Sprat were aware of the fact that poetic wit consists in perceiving "striking resemblances between things," the difference between poetry and other types of writings was largely viewed in terms of meter and "poetic license."[7] Lacking theoretical support, new developments in poetry were soon eclipsed by a poetry, which excelled in intellectual argumentation and logical coherence. Poetry turned into what Roland Barthes refers to as "classic texts," that is, texts which limit the constructive freedom of their interpreters by reserving for the artist the complete "mastery over meaning" (Barthes 1974: 174). With poetry moving closer to scientific discourse, the question of its uniqueness was soon to be raised. It is at that moment that the poetic achievements of the beginning of the century loomed large, conveying to the understanding that the "language of the passions" is different from that in which "reason speaks" (Hamilton 1963: 41).

Whereas music reflects new intellectual conditions only indirectly, primarily through artistic changes, literature also partakes directly in the debates over new queries and outlook.[8] The literature of the first half of the seventeenth century made an unerring impression on the conditions to which it responded. "The new philosophy calls all in doubt," wrote John Donne in his "Anatomie of the World."[9] Doubt was cast upon responsibility and freedom, and upon all knowledge inferred by procedures other than those provided by the "new method." We have already seen this suspicious attitude reflected in theories of expression, but poets were confronted by a factor that had knowingly been avoided, that is doubts concerning the status of language (of the kind we have encountered in Mersenne). By incorporating the predicament of language within their doubts, the English poets paradoxically succeeded in opening new avenues for poetry, poetics, and the philosophy of art.

The problem of language was a direct implication of the new method. Words had a hard time finding their place in a "book written in mathematical symbols." In Galileo's "book of the world," words not only ceased to be essences in themselves, as was previously held, but could not even refer to that which was considered essential, namely, the Cartesian "primary qualities" related to the physical structure of nature. Yet, the world of "secondary qualities"—of experience, especially of the passions—could not be ignored. Words and "second qualities" were thus destined to collaborate, albeit in an unfavorable intellectual climate. Their new status called for a redefinition of their relationship. Above all, it required the realization that the secondary qualities, unlike the primary, are elusive and liable to perpetual change. The power of language to express while creating, to realize that to which it refers, required, accordingly, careful examination.

Though aware of their own linguistic powers, poets were not as determined with regard to their specific undertaking as were contemporary musicians in Italy, who were able to clear the stage for an all encompassing new enterprise. In poetry, no clear discrimination was made between old and new uses of the arts. Poets, moreover, had to face a verbal discourse, which gradually swallowed the discourse they themselves were after. For Bacon and Hobbes, for example, ordinary language was "laced with misleading qualitative judgements" wrongly assumed to belong to the objects observed (Hamilton 1963: 13). Though in and of itself not exactly a scientific tool, language, it was argued later in the century, may serve, nonetheless, for the description of method and procedure, and the tracing of assumptions, arguments and the like.[10] Emphasis was placed, therefore, on simple, clear, and unambiguous language. Under the influence of the discussions held at the Royal Academy, this kind of language was adopted not only for pure scientific accounts, but also for all kinds of "useful" discussions. By the end of the seventeenth century, it infected even the language of poetry, which tried to rid itself of the eloquent exuberance of former poets (Hamilton 1963).

The Changing Status of Metaphor

But the situation was different in the course of the first decades of the seventeenth century. While changes were in the air they were not yet fixed by either method or values. Seeking a new orientation, writers substituted the old Ciceronian norms by the Senecean prose style. "Their purpose of writing," writes Croll, "was to portray, not a thought, but a mind thinking, or, in Pascal's words, *la peinture de la pensée*. They knew that an idea separated from the act of experiencing is not the idea that was experienced. The ardor of its conception in the mind is a necessary part of its truth; and unless it can be conveyed to another mind in something of the form of its occurrence, either it has changed into some other idea or has ceased to have any existence whatever except a verbal one" (Croll 1926).

As with prose, so with poetry: poets at the beginning of the seventeenth century searched for a language which would *capture* the occurrence of the experience itself. Verging on an almost Humean conception of language, it should be emphasized again that these writers were not interested in theory; their orientation was mainly practical. Like the founders of the opera, English poets were defining powers in the process of creating them; like them they tried to establish a new paradigm for their art. As in the case of opera, their new paradigm reflected the paradigmatic change in science. It was built on the assumption that reality—physical or psychological—is to be directly observed and manipulated.

But unlike the musicians of the period the so-called metaphysical poets did not suggest an altogether new artistic form such as the one associated with accompanied monody. Yet, like their fellow musicians, they regarded rhetoric as their main source of inspiration. Attracted by its cognitive rather than by its instructive or decorative aspects, they elaborated all kinds of *figurae sensum,* within a large scale of intricate constructions, for the sake of activating thought.[11]

But how do such figures activate thought? Present day philosophy and psychology may be of help. From the road paved by Richards (1936), Black may have been one of the first (in the last decades) to have developed this notion to the point at which metaphor ceased to be considered as a mere eloquential device (Black 1962).[12] Accordingly, Black treated the *figurae sensum* cognitively, rather than rhetorically, replacing "metonymy," "synecdoche" and other traditional figures by metaphors of various kinds—"substitution," "comparison," and "interaction." The first two kinds of metaphor, argued Black, account for decorative figures of speech: "substitution" includes all cases in which a literal expression can be simply replaced by its metaphoric equivalent; "comparison" provides a more elaborate paraphrase of the relation between the subject and its metaphorical predicate. "Interaction," explains Black, uses words not as lexical indices but as systems of "associated commonplaces" in which one element relates to the "frame"—the sentence in which the metaphor occurs—and the other to the, so-called, "focus"—the word which is used metaphorically. Interaction metaphors have the power to "select, emphasize, suppress and organize features of the principle subject by implying statements about it that normally apply to the subsidiary subject" (ibid. 44-5). Whereas substitution metaphors and comparison metaphors, according to Black, can be replaced by literal translations, with no loss of cognitive content, the third type of metaphor implies a distinctive intellectual operation, it requires "simultaneous awareness of both subjects not reducible to any comparison between the two."[13] Hence, only interactive metaphors are of real philosophical interest, according to Black, insisting that it is impossible to provide a literal paraphrase of interactive metaphors, for it "fails to give the insight that the metaphor gives." Metaphor is always a one time success which no rules can determine apriori. Indeed, a good metaphor, psychologists agree, is the product of a *creative act*, which has, at one and the same time, informative value, hence its relevance not only to artistic creativity, but to the theory of knowledge as well (Brunner 1965; Sacks 1979; McCormac 1985, 1986; Miall 1986; Dent 1987; Fainsilber & Ortony 1987).

Though poetry at all times has capitalized on metaphors, it might be argued that compared with their predecessors, the "metaphysical" poets used metaphor in a much more "interactive" way, for they employed the new resources of the conceit which consisted of juxtaposition of images and comparisons between very dissimilar objects. Moreover, many of their poems enlist metaphors interacting with each other. The resultant complexity defies simple analogical relationship; metaphor and reality can be reconciled only in imaginative terms. Using these criteria, examine Donne's *A Valediction: Forbidding Mourning* (textual example 2.1):

Textual Example 2.1

As virtuous men pass mildly away,
And whisper to their souls, to go,
Whilst some of their sad friends do say:
'The breath goes now', and some say: 'No':

So let us melt, and make no noise,
No tear-floods, nor sigh-tempests move;
'Twere profanation of our joys
To tell the laity our love.

Moving of the earth brings harms and fears'
Men reckon what it did and meant:
But trepidation of the spheres,
Though greater far, is innocent.

Dull sublunary lover's love
(Whose soul is sense) cannot admit
Absence, because it doth remove
Those things which elemented it.

But we, by a love so much refin'd
That our selves know not what it is,
Inter-assured of the mind,
Care less, eyes, lips, and hands to miss.

Our two souls therefore, which are one,
Though I must go, endure not yet,

A breach, but an expansion,
Like gold to airy thinness beat.

If they be two, they are two so
As stiff twin compasses are two:
Thy soul, the fix'd foot, makes no show
To move, but doth, if the other do;

And though it in the centre sit,
Yet when the other far doth roam,
It leans, and hearkens after it,
And grows erect, as that comes home.

Such wilt thou be to me, who must,
Like the other foot, obliquely run:
Thy firmness makes my circle just,
And makes me end where I begun.

As we can readily see, the title of the poem posits a dissonance that calls for a solution in the unfolding of the poem. The metaphorical construction of the poem is directed toward that goal: earthly separation finds its justification, as the poem concludes, in a spiritual reunification. Obviously this calls for transformations in the course of the poem: The poem exploits four basic metaphors, presenting complicated relations between "frame" and "focus." Both "frame" and "focus" pinpoint the direction of the interaction, leaving ample room for the reader's interpretation of each metaphorical construction and the coalescence of all four. Indeed, as one examines the four metaphors in succession, one becomes increasingly aware of the transformation that had taken place from earthly "valediction" to a philosophical reaffirmation of steadfast union, capable to withstand "trepidations."

As exemplified in Donne's poem, which combines the glory of the spheres with the mundane of a compass, poets in the seventeenth century subjected realistic objects, no less than imaginary ones, to metaphorical constructions. Within the boundaries of a poem all images were liable to be treated metaphorically. A comprehensive treatment of this magnitude cannot be ascribed solely to the changed status of certain symbols, like those related to the harmony of the spheres; rather it points to a major change in the conception of knowledge. By appealing to new cognitive resources, the metaphysical poets paved new ways to convey the changing conditions of emotional life, as such, they became a model even for modern poetry.

The Shift of Attention to the Nonreferential Aspects of Poetry

Up to the sixteenth century "resemblance" played a constructive role in the theory of knowledge of Western culture. It was resemblance, explains Foucault, that "largely guided the exegesis and interpretation of texts; it was resemblance that organized the play of symbols, made possible knowledge of things visible and invisible, and controlled the art of representing them. The universe was folded in upon itself: the earth echoing the sky, faces seeing themselves reflected in the stars, and plants holding within their stems the secrets that were of use to man" (Foucault 1972: 17). By contrast, the changes brought about by the new science emphasized new types of observation and experimentation, governed by attempts to unveil the regularity of cause and effect. Knowledge, liable to refutation and verification, is inconceivable without that change. Inferences could no longer be drawn from subjective judgments than could universal criteria be based on mere belief.

Placing "resemblance" outside the boundaries of science, the separation of reality, fantasy and myth became more clear-cut (as the Fludd-Kepler debate so beautifully exemplifies; see chapter 1, fn. 30). With regard to reality, one looked for the closest relationship between cause and effect, and tried to embed common elements in measurable units. While objectivity was thus guaranteed, the world of knowledge seemed, nonetheless, impoverished with the decline of *convenientia*, *aemulatio*, *analogia* and the play of *sympathy*.[14] Poets seem to have filled this gap by enlisting resemblances as their main equipment. Investigating new ways of activating resemblances, poets employed procedures that could compete with either scientific truth or religious dogmatism, offering not only psychological and cultural satisfaction, but intellectual satisfaction as well. By means of resemblances forged as metaphors, they were able to cope with new existential conditions while securing the continuity of intellectual habits and contents.

The attempt to come to grips with poetic thinking was related not only to analogical thinking, but to many of its implications as well. Indeed, what was inaugurated by literary men of the first half of the seventeenth century called attention to the nonreferential aspects of poetry. Having considered the latter as a kind of "musicalization," Hollander (1956) showed great insight into a latent process that ended with a full awareness of the nonsemantic in poetry. With the "suspicion of poetry of her own features," he writes, "poetry, staring and gazing at her musical sister, came to see in that vision more and more of itself." While this characterized seventeenth-century English poetry in general, in his *Untuning of the Sky* (1961) Hollander also deals, as was mentioned above, with a more concrete musical trend in poetry, one which employs myths and metaphors related to the powers of music. By assigning language the ability to *activate* those metaphors and to vivify old myths, it heralded the newly defined powers of language. Indeed, the "musical" trend in poetry, which transformed world harmony into poetical content concerning the powers of music, revealed, paradoxically, another way of inciting the intellectual faculties. In the debate over what in music affects the hearer—harmonious proportions or linguistic properties—the latter, undoubtedly, had the upper hand. This trend, in other words, contributed to the renewed identification of music with rhetoric. It linked music and poetry via metaphor, even though the metaphors were *about* the powers of music, rather than the outcome of the interaction *between* poetry and music.

The following verses by Campion (textual example 2.2) say it all, but more poetically. Notice how his musical setting (musical example 2.1) supports the text without claiming independence:

Textual Example 2.2

Awake thou spring of speaking grace
Mute rest becomes not thee;
The fayrest women, while they sleepe,
And Pictures equall bee.
    O come and dwell in love discourses,
    Old renuing, new creating.
    The words which thy rich tongue discourses
    Are not of the common rating.

Thy voyce is as an Echo cleare
Which Musicke doth beget,
Thy speech is as an Oracle
Which none can counterfeit;

For thou alone, without offending,
Hast obtain'd power of enchanting;
And I could heare thee without ending,
Other comfort never wanting.

Some little reason brutish lives
With human glory share;
But language is our proper grace,
From which they sever'd are.

As brutes in reason man surpasses,
Men in speech excell each other;
If speech be then the best of graces,
Doe it not in slumber smother.

Musical Example 2.1
"Awake thou Spring"

From: Thomas Campion, *Third Book of Ayers*, 1617, ed. by Edmund Horace Followers, Stainer & Bell, 1925, pp. 24-5.

Musical Example 2.1 continued

That Thomas Campion (1567-1620) was a poet-musician is evident from the sheer sound of his poem. Nevertheless, he grants "grace" to speech and language at a time when such attributes were more readily bestowed upon music. His literal praise of language emphasizes its ability to reason and highlight the superiority of man among living creatures. He set the poem to music in such a way that it supports the text without claiming its own independence, "clear like an echo" it is strophic, syllabic, monodic, melodious without being affective. As far as he was concerned, the place he granted music did not contradict his Platonic view that harmonious relations should govern the prosodic elements of the verse, as will be discussed later in this chapter.

The Powers of Music as an Effective Bridge

Mace claimed long ago (1952), that the new philosophy in seventeenth-century England delivered a deathblow to the speculative considerations of music. Given this to be the case, then poetry, as suggested by Hollander, has turned into a legitimate area for new explorations. Eighteenth-century theoretical interest in musical cognition, we maintain, is related to these earlier processes in ways more intimate than one would have supposed. Continuously in search of images better suited to contemporary beliefs about music, the poetry, which dealt with the powers of music, served as a bridge between the old metaphysics and the forthcoming aesthetics. The poetic elaboration, for example, of images related to the "tuning of the soul," led to a final secularization of the notion. Related sub-images, such as the tension of the hearer or the proper temperament of either soul or body, eventually converged in the "psychic tonus" that was to be a central concept in later theories.

The blend of the actual and the metaphysical is particularly revealed in Milton, who was aware of the fact that the "heavenly harmony of voice and verse" is a consciously *contrived* process (Hollander 1961: 330). The convergence of the actual and the metaphysical, unique to the artistic realm, occurred, interestingly, at the same time in poetry, painting, and music. In poetry and painting the mythological lyre could be transformed into a modern harpsichord; muses such as Euterpe and Erato, later depicted as simple Dutch maidens, could play modern instruments (pictorial example 2.2). In music, praying was transformed into singing (Pirrota and Povoledo 1982) and "touches of celestial harmonies" into practical musicianship. Though the intellectual climate has changed, artists, unrestricted by theoretical constrains, could freely mingle images, ideas, and practices of various kinds, aware that the worlds they created better reflect their experiences and thought.

What started as metaphorical elaboration, eventually turned into the construction of models explaining the powers of music. Augustan poetry, says Hollander, "provides more of a narrative logic for the introduction of the notion of heavenly music than relying heavily on a turn of wit, as in both Metaphysical and Jonsonian traditions" (Hollander 1961: 356). Practically oriented, the narrative it tried to provide became more and more like an account of real psychological processes. At the turn of the seventeenth century, before the complete takeover of the theoretical orientation, we witness an overlap between musical imagery in poetry and theories related to the expressivity of music. The importance of the antecedent stage goes beyond the transmission of ideas, for it provided models and metaphors, which were eventually to govern major theoretical breakthroughs. "Movement," "sympathy" and "vibration" turned into metaphors, playing as such central roles in later theories, as we shall yet see. The ideational content of music, which was denied by the new science, could by then be transformed from the imaginary position it held in poetry to an assumed fact that could be accounted for in psycho-logical terms. Strode's poem, *In commendation of Musick* (textual example 2.3, quoted from Hollander 1961: 295) though it belongs to the first half of the seventeenth century, calls attention to the role of poetry in the transition from a philosophy that denied music's harmonious power to that which will reify it albeit on a different basis.

Textual Example 2.3

When whispering straynes do softly steale
With creeping passion through the hart,
And when at every touch wee feele
Our pulses beate and beare a part;
    When threads can make
    A harstring shake
    Philosophie
    Can scarce deny
The soule consists of harmony.

When unto heavenly joy wee feyne
Whatere the soule affecteth most,
Which onely thus wee can explayne
By musick of the winged hoast,
        Whose layes wee think
        Make starres to winke,
        Philosophie
        Can scarce deny
Our soules consists of harmony

O loll mee, lull mee, charming ayre,
My senses rock with wonder sweete;
Like snow on wooll thy fallings are,
Soft, like a spiritts, are thy feete:
        Griefe who need feare
        That hath an eare?
        Down lett him lye
        And slumbering dye,
And change his soule for harmony.

Note that the poem relies heavily on the *experience* of music, rather than on collective myth, thus presenting a challenge to those who separated mind from heart and the sensuous from the affective. While still supported by "the winged hoast" to explain the joy of music, the poet calls upon sympathy based on synaesthetic images, rhythmic activation, and the like, fully convinced that these earthly matters are crucial manifestations of the power of music.

In the Twilight of *Ut Pictura Poesis*[15]

Some of the major stages in the career of the *ut pictura poesis* tradition in the English literary pictorialism coincided almost exactly with the attempt, described above, to "redefine the powers of language." The impact of this venerable tradition seems to have diminished at the beginning of the seventeenth century, to have grown towards its end, and gradually to have lost its power from the middle of the eighteenth century onward. Here, too, theory was lagging behind, for it did not discern these changes, let alone provide an explication. For that matter, the same premises and assumptions appear to have underlined different and even contrasting kinds of pictorialism. Only toward the end of the seventeenth century, one may find some significant connection between theory and practice, when dogmatic formulations of the former seem to reflect and even to encourage manneristic approaches in the latter. As in the phenomenon of painting borrowing from poetry, there were two major principles that guided poets in applying the *ut pictura poesis* premises to the organization of poetry along pictorial models: *enargeia* and *exphrasis*. The two principles involved artistic procedures not initially inherent in the linguistic medium. The first, concerning the vividness of the description, was an attempt to evoke the *mind's* eye through verbal stimuli; the effects were supposed to come as close as possible to the effects obtained by the sensual eye, thereby achieving the desired illusion. The second principle, revealed in the attempts to render verbally real or imaginary art objects, created its own genres, referred to sometimes as iconic poetry and prose.

The two principles were not mutually exclusive; they could complement each other. As far as iconic poetry was concerned, the textual unfolding also tried to capture the movement of the eye perceiving the objects; *enargeia* could thus be obtained through a more definite point of view than the one provided by nature. Art works, in fact, supplied literature with organizing models that could be followed either meticulously or freely. In other words, Poetry could approach painting and sculpture by turning directly to the objects it treated or by borrowing from her sister arts some of their structural procedures. English poets of the high Renaissance employed both alternatives, but it was primarily "*enargeia*, or clearness of representation," as George Chapman put it, which got most of the attention; it was the intensification of the *perceived* that was mainly desired (Hagstrum 1958: 63). Considering the centrality of illusion in Renaissance art, this should not come as a surprise.[16]

However, at the beginning of the seventeenth century a change occurred, which bestows upon iconic poetry a new role. Though iconic devices provided poetry with structural elements that were neither narrative nor discursive, they were now less employed for descriptive purposes. Thus, for example, the poetic "gallery," as Hagstrum called it, organized poetical sections as a series of tableaux, which attempted to weave composite cultural images, rather than descriptive or natural scenes. The following stanzas from Marvell's *Gallery* beautifully illustrate this tendency (textual example 2.4):

Textual Example 2.4

I

Clora come view my Soul, and tell
Whether I have contrive'd it well.
Now all its several lodgings lye
Compose'd into one Gallery;
And the great Arras-hangings, made
Of various Faces, by are laid:
That, for all furniture, you'l find
Only your picture in my Mind.

II

Here Thou are painted in the Dress
Of an Inhumane Murtheress;
Examining upon our Hearts
Thy fertile Shop of cruel Arts;
Engines more keen that ever yet
Adorned Tyrants Cabinet;
Of which the most tormenting are
Black eyes, red Lips, and curled Hair.

III

But, on the other side, th'art drawn
Like to *Aurora* in the Dawn
When in the East she slumbr'ing lyes.
And stretches out her milky Thighs;
While all the morning Quire does sing
And Manna falls, and Roses spring;
And, at thy Feet, the wooing Doves
Sit perfecting their harmless Loves.
[...]

VI

These Pictures and a thousand more,
Of Thee, my Gallery do store;
In all the Forms thou can'st invent
Either to please me, or torment:
For thou alone to people me,
Art grown a num'rous Colony;
An a Collection choicer far
Then or *White-hall's* or *Mantua's* were.

VII

But, of these Pictures and the rest,
That at the Entrance likes me best;

> Where the same Posture, and the Look
> Remains, with which I first was took.
> A tender Shepherdess, whose Hair
> Hangs loosely playing in the Air,
> Transplanting Flow'rs from the green Hill,
> To crown her head, and Bosome fill.

The images of this poem, though not necessarily complementary, are viewed from a unified psychological vista—a lover's point of view. The iconic succession does not distort the overall emotional intent, quite the contrary, it is subjugated to the same creative enterprise which the above-mentioned contemporary painters, who dealt with expression, undertook. Indeed, despite of itself, even this kind of iconic poetry seems to have contributed towards the replacement of *pictura* by *musica.*

No statement could better capture the change that has taken place in the seventeenth century than Marino's reformulation of the Simonidean antinomy according to which painting is mute poetry and poetry is a spoken picture. Whereas painting "causes us almost to understand with the senses," tells us Marino, poetry makes us "feel with the intellect." According to Marino, the intellectual in painting is not confined to narrative contents, nor is the sensual in poetry restricted to rendering of the visual. Poetic feelings now seemed to comprise all senses, above all, those created by the movements "of the sensitive parts of the soul"—the emotions, according to Descartes. As for the visual, the "innocent eye" of the Renaissance was replaced by an eye equipped with various kinds of artistic sensibilities; the definition of the pictorial in poetry, thus, became rather evasive.

While poets in the first half of the seventeenth century were preoccupied with visual images in an evocative capacity, those of the neoclassical period (the second half of the seventeenth century and later) were busy with much stricter modeling. The modeled picture in such poetry became the matrix determining the details of the poem; resting on narrative and historical paintings, the condensed plot of the picture, became, in fact, the focus of the borrowing. The work as a whole exploited an "equation" that worked for the total argument, as well as for its separate segments. At the same time, the work comprised a piece of reality, well known to contemporary observers, vividly displaced and easily imagined.[17] One may discern this tendency in the following part of Thomson's

Seasons (ll. 113-29; textual example 2.5), which took Guido Reni's *Aurora* as its model (Hagstrum 1958: 260). In order to view how meticulously Thomson observed the picture and restaged it poetically and to what extent he limited himself to the pictorial scene, the reader is advised to look at pictorial example 2.3 which presents Reni's picture.

Textual Example 2.5

Parent of Seasons! who the pomp precede
That waits thy throne, as through thy vast domain,
Annual, along the bright ecliptic road
In world-rejoicing state it moves sublime.
Meantime the expecting nations, circled gay
With all the various tribes of foodful earth,
implore thy bounty, or send grateful up
A common hymn: while, round thy beaming car,
High-seen, the Seasons lead, in sprightly dance
Harmonious knit, the rosy-fingered hours,
The zephyrs floating loose, the timely rains,
Of bloom ethereal the light-footed dews,
And, softened into joy, the surly storms.
These, in successive turn, with lavish hand
Shower every beauty, every fragrance shower,
Herbs, flowers, and fruits; till kindling at thy touch,
From land to land is flushed the vernal year.

The imagination at work in this sort of pictorialism was limited to "reconstructions." The pictorial reality, remade through the objects exploited by poetry, determined not only the domains to which the poetic images were applicable, but also the ways in which they were to be used. The "associated commonplaces" the picture brought along was directly applied to the poetic domain, but did not interact with it; rather, the narrative structure coincided with an organization already otherwise "effected in the new realm," as Goodman (1968) would have claimed. In other words, the organization of the picture was poetical to begin with, using as it did allegory and myth, and once used "became routine."

However, under the auspices of the *ut pictura poesis* tenets, which still dominated the scene at the beginning of the eighteenth century, poetry turned more and more to *fluctuating sentiments*, looking for

Pictorial Example 2.3
Guido Reni: Aurora

pictorial elements that might yield "air" and "atmosphere." Commenting on Collins, for example, Hagstrum says "he draws a gradual dusky veil over the real world so that we may not be too long diverted from our exploration of the 'World Soul' and 'the dim-discover'd Tracts of the Mind'" (Hagstrum 1958: 286). At the same time, *temporal* unfolding became even more prominent than "atmosphere" that overrode visual details. Poets, Barry (1987) emphasizes, were not only preoccupied with synaesthetic images, but with narrative structures, dramatic tensions and the like. Their predilection for a mixture of pictorial and musical images may have suggested the poetic tendency towards an aesthetic and cognitive alternative.

Thus toward the middle of the eighteenth century, poetry seemed to be somewhat suspicious of pictorialism; yet poets could not ignore the profits gained thereby. One way to overcome the suspicion and still enjoy pictorialism was to divide poetry into a "musical" and a "pictorial" kind. "Lyrical poetry," writes Hildebrand Jacob in 1734, "approaches more to Music... as dramatic and pastoral poetry do to Painting."[18] Whether or not his statement withstands critical examination, it clearly testifies to an attempt to break the shackles of the *ut pictura poesis* doctrine on a theoretical level.

On second thought, it is no surprise to find romantic poets to have "repudiated the notions that poetry should make pictures with words" insisting on its "intellectual freedom from visible form." Nor is it surprising to find romantic painters resenting the expectation that they "could illustrate poetry" [or] should represent the "pregnant moment" of a poetic or historical narrative that literature alone could explain "in full" (Heffernen 1985: xviii). What the romantic poets were after, and in what way they still used pictorial elements is, of course, beyond the scope of the present study.

Animating Time: New Rhythmical Explorations

Poetic Measure and Rhyme: A Reappraisal of Their Function

While the concentration on the powers of music was limited to occasional poetry, experiments with "real" music—the prosodic aspects of the text and its overall rhythmical structure, preoccupied poets throughout the seventeenth century. Toward the end of the sixteenth century, a new interest arose in the rhythmical parameter of the text and its theoretical implications. This was a period in which the blank verse declined after having achieved perfection in Shakespeare. For most Elizabethan poets rhythmic procedures became strict or simply conventional whereas their lyrics were somewhat more successful (Saintsbury 1923 ii: 100).

In search of "musicality," poets turned to music, which had already proved itself by "suggesting fairly complicated but harmoniously concerted measures." Yet poets frequently confused musical and poetic music, "failing to achieve the latter in their anxiety to suit the former," writes Saintsbury.[19] Thus, metrical confusion became mixed with the general doubts concerning the poetic qualities of English. Waging war against the practice of "barbarous riming," poets tried to apply classical quantitative meters to English. A new impetus was gained with the poet-musician Thomas Campion, quoted above, who, like St. Augustine, considered quantitative prosody a temporal embodiment of the *harmonia mundi*. The incompatibility between theory (that hailed quantitative and unrhymed prosody) and practice (which used stress and rhyme) disturbed neither him nor his friends.

In 1603, Samuel Daniel contradicted the "new paradigm" suggested by Campion and his friends. His own ideas were advanced in order to justify old practices by giving them a "modern" assessment. For him rhyme was still a major poetic procedure, for it gave "both to the Ear an echo of a delightful report, and to the Memorie a deeper impression of what is deliverd therein" (Daniel 1930: 132); notice that the figure of echo also appears in Campion's song mentioned above. For Daniel, as for Vincenzo Galilei, harmonious numbers no longer had artistic significance. Artistic value, Daniel implied, is determined experientially; ear and intellect, he added, in the spirit of Aristoxenus, are its sole judges. Yet, unlike Galilei, Daniel believed in a separation of sound patterns and their message: Effective as "Number, Measure, and Rhyme" may be, they are but "as the ground or seate, whereupon is raised the work that commends it." Sound and meaning are related to each other only causally; the nonsemantic parameters function as "deliverers," facilitating the road to understanding.[20] Acknowledging the importance of cultural context, Daniel did not look for elements of universal appeal in prosody. Psychological factors, he implied, were embedded in habits qua habits rather than in their specific content. Nonetheless, his analysis touches upon issues that will prove essential to the bridging of subject and object, sense and mind. His observation concerning the relevance of rhyme to coherence and memory reveals an initial awareness of the fact that the perception of the musical elements of poetry involves cognitive aspects. At any rate, such observations freed poetics from theoretical prejudices enabling poets to try out new artistic possibilities.

Indeed, with Donne at one extreme, Dryden and Pope at the other, and Milton in between, wide divergences in rhythmic norms may be discerned. At first glance it seems that these different kinds of verse have nothing in common. Yet it is the differences, which highlight what they have in common, namely, a correlation between the world of meaning engaged by the poet and the kind of rhythm he employs. Sound, as Pope advocated (echoing both Daniel and Campion), became the "echo" of sense in multiple ways. Daniel's contention that the rhythmical design should help deliver the message was thus applied in ways beyond those he had envisaged.

Form and content were already intimately associated by the Spenserian poets, yet the "harmony" achieved was an outcome of an actual connection with music. "The effect of writing with music in view," observed Cox, "was to preclude complexity of mood, rapid shifts of tone or a freedom of movement suggesting the speaking voice" (Cox 1956: 48). Bringing poetry to the province of music, entailed the sacrifice of poetry's own complexity as a literary art. In Italy, a similar tendency led to a real "invitation" to music that resulted in the *drama per musica*. The musicalized verses of Tasso and Guarini were transformed by Striggio and Rinuccini into opera libretti. In England, once music had lost its metaphysical status, poetry was granted full autonomy; words, preferred by the English at all times, gained total predominance. Donne, for example, though perhaps still intending part of his poems for music, seldom allowed music "to control their diction, movement or form much less their theme, attitude, or general development"(Cox 1956: 49). In that sense, the break of seventeenth-century English poetry from real music, parallels the ascendancy of instrumental music, which marked the emancipation of music from textual dictates.[21] The increased autonomy of each of the arts opened up a new region for artistic explorations—the region of *conscious* metaphorical interactions among the arts.

Music Freeing Poetic Content from the Fetters of Verse

Trying to bring music to poetry instead of poetry to music, poets were now trying to wed the music of verse to its content. Donne's general spirit of experimentation, for example, penetrated his metrics as well; Donne, writes Grierson, is "one of the first masters.. of the elaborate stanza or paragraph in which the discords of individual lines or phrases resolved in the complex and rhetorically effective harmony of the whole group of lines."[22] The performance of the poem became a challenge to the reader; he was called upon to enlist his powers in unexpected ways. The general tone of the poem was also defined; it was direct and personal, granting verse the air of living speech, rather than that of formal and public utterance.[23]

Read, for example, the following poem, *The Triple Fool* (textual example 2.6, quoted from Donne 1952: 14), which is interesting for its self-referentiality with regard to its prosodic qualities. In "performing" the poem, the readers should be alert to the uneven distribution of the stresses. They are called upon both to contract as well as lengthen some of the syllables in order to create the desired balance among poetical segments respectful of their syntactic, rhetorical and semantic constraints. This is made possible by the fact that the poem is primarily based on monosyllables, which may be equally stressed or unstressed. This, Webb tells us, is the prime condition for musicalizing poetry, for it makes the rendering of the fluctuations of emotions possible (see Vol. II, p. 287ff.). Indeed, the power of performance of such a poem frees it from "the fetters of verse" injecting the affect, which has become poetically "fixed," with new emotive life.

Textual Example 2.6

I am two fools, I know
For loving, and for saying so
In whining Poëtry;
But where's that wiseman, that would not be I,
If she would not deny?
Then, as the earth's inward narrow crooked lanes
Do purge sea-water's fretful salt away,
I thought, if I could draw my pains
Through rhyme's vexation, I should them allay.

Grief brought to numbers cannot be so fierce,
For he tames it, that fetters it in verse.

But when I have done so,
Some man, his art and voice to show,
Doth Set and sing my pain,
And, by delighting many, frees againe,
Griefe, which verse did restraine.
To Love and Grief tribute of Verse belongs,
But not of such as pleases when 'tis read;
Both are increased by such songs:
For both their triumphs so are publishéd,
And I, which was two fools, do so grow three;
Who are a little wise, the best fools bee.

Intimacy of mood and an intimate relationship between text and its rhythmical texture were prime concerns of the metaphysical poets, as well as those associated with the school of Jonson. As expected, with the Restoration, emphasis was put more on political poetry, for which oratory seemed the most appropriate model. The variety of artistic expressions that centered on private affairs and intimate experiences gave way to a uniform style that was concerned with the general experience of the public. However, the search for "appropriate music" did continue. The ascendancy of an "oratorical" poetry went hand in hand with the struggle for the "heroic couplet." A well-defined world of meaning sought expression in a regular and unambiguous rhythm that would not clash with the rhetorical delivery of the text. Mechanical procedures—intelligible, repeated patterns that guaranteed clarity—were substituted for the rhythmic complexity that marked the earlier style. Once the "heroic couplet" was turned into an inflexible patent form, they could no longer "reach meaning through ways not distinctly traceable by the understanding," as Pater would have claimed (Pater 1980: 108). Nevertheless, they survived for a hundred years almost unchallenged.

It is against this hegemony that a man like Daniel Webb, a key figure among our British writers, later revolted. His ideal, however, was that of Milton rather than that of Donne. Of course, he was not alone in that preference; by the second half of the eighteenth century, Milton's poetry had already been recognized as the height of the achievement in musicalized verse, a recognition that holds to this very day. For Milton, says Hollander (1961), "the new separation between the experience of actual music as one of life's various phenomena and the response to a philosophy of music is entirely arbitrary." In poems praising the powers of music like "At a Solemn Musick," for example, Milton vivifies the myth and the metaphors he uses not only ideationally but also by dressing them in the proper musical attire. According to Finney (1962), Milton, in his later works, was ahead of his time. He turned away from metaphysical ideas toward a musicalization of his poems by borrowing from the contemporary Italian musical drama structural and dynamic elements.

Less personal than Donne, Milton avoided being caught in oratory so typical of later poets, granting his mythical subjects the grandeur characteristic of an epic style. Whereas his rhythm lacks the immediacy and intimacy characterizing the metaphysical poets, his verse is seldom harsh or rigid. Euphony is at all times his prime concern, concordant with his aims to resolve all "undiscording voices" into "melodious noise." Again, we find compatibility between content and form, that is, between "nature and destiny of man" and the "magnificent march" in which it is carried through.[24] If throughout the whole century poetry aimed to wed form and content—what was later considered as a real aspiration toward the condition of music—why did Milton secure the position of *the* "master of harmonies" in English poetry? The attribution of musicality to Milton seems to imply more than either Pater, Spitzer, or Hollander entertained, for it concerns the organization of rhythm and the handling of sound according to prevalent musical examples, emphasizing the power of poetry.

Controlling the Flow of Time in Music and Poetry

In chapter 1 we maintained that the real challenge of musicians in the Middle Ages was to achieve control over the flow of time. When this was attained, we argued, the autonomy of music was guaranteed. Once the musical beat was engaged as the basic unit of coextensive hierarchical divisions, textual aid was no longer required for "syntactic" coherence. Variety and flexibility became possible within simultaneous occurrences without sacrificing coherence. The conception that music is rhythm-oriented prose and that its rhythmical segments fall into unfixed and not necessarily repeated patterns (despite poetical procedures of the classical style) rests largely on these rudimentary facts.

In comparison with musical rhythm, poetry addresses the flow of time only "tacitly," to use Michael Polanyi's apt term[25]; the unfolding of meaning remains its main focus, even though it is embedded in rhythmical structures. Consequently, the musical parameters in poetry are less defined; their "notation" is ambiguous and hidden within a verbal system. It is thus difficult to compare the arts, even in this shared physical dimension. In the continuous debate over the components of prosody and their interplay, there is, however, a general agreement over one property that poetry and music share: Poetry, like music, comprises a "heightened and strongly directed continuity among the various segments," as Kramer (1984: 8) puts it, resulting from a "physical pulse" marking time.

What was accomplished in music towards the end of the Middle Ages took place in poetry by the time of Milton, though it could not appear in the foreground. It was made possible through an interaction of poetry and music; the interaction seems to have occurred at the moment when polyphonic music, with its continuous rhythmical flow, had achieved perfection. Once recognized as music at its purest, its rhythm became identified with what we understand by "musical time"—time experienced in accordance with the "manipulation of its passage" (Langer 1953). But it is only after the madrigalists subjected this continuity to rhetorical needs that this procedure became available to poets. It is their freedom of phrasing, along with the continuous motion concomitant with the general spirit of the piece, that became translatable into poetic structures.

Marenzio's madrigal (musical example 2.2) demonstrates these qualities beautifully. In this phase in the development of the madrigal we witness a shift from a unified homophonic texture and an even, homogenous pace, where all sentences are more or less equally treated, to a highly differentiated treatment of individual poetical lines. By exploiting the rich vocabulary of musical devices, available at the time, the late madrigal attempted to impart the semantic content of a few key words, which dominate each of the lines.

The above excerpt, which is a highly sophisticated rendition of Petrarch's famous sonnet, *Solo e pensoso,* may be divided into two major sections, each of which may be further subdivided. The first

Musical Example 2.2
Luca Marenzio: *"Solo e pensoso"*

Luca Marenzio, Il nono libro de magrigali a 5 voci (Venice: Gardano, 1599), ed. Iain Fenlon, Music and Patronage in Sixteenth-Century Mantua (Cambridge: Cambridge University Press, 1982) 2:99-105.

Musical Example 2.2 continued

Musical Example 2.2 continued

Musical Example 2.2 continued

segment of the first section is dominated by the words: *solo e pensoso* (alone and pensive) while the second segment of the first section is dominated by the words *tardi e lenti* (retardant and slow). The two segments of the first section create a contrast between a static movement and a halt. The static movement is achieved by motion and counter-motion, which cancel each other, underlined by a nondirectional harmonic progression. The second segment is gradually ushered in, and is connected to the first one through the chromatic progression of the soprano, which creates an overriding binding arch. The total halting is effected by slow note values, which are almost equally distributed. An added *supplimento,* as Burmeister would have called it (in the alto and the tenor), helps create "the sense of an ending." By contrast, the first segment of the second section establishes a totally new kind of movement—that of *fuggire* (fleeing) through a "fugal" imitative technique exploiting new note values which generate a syncopated rhythm. This restless segment is contrasted by an even pace of the quarter notes—the steadiness of the *"human...stampi"* (human footsteps).

As Burmeister insightfully stated already in 1607, each such sentence (i.e., a musical "point") functions as an "affect" unto itself. Whereas opera welcomed the emphasis on affect, it could not afford, from a mimetic point of view, to accept the breaking up of affect into isolated, contrasting statements of the kind exemplified in Marenzio's poem. Poetry, on the other hand, made the most of it. As Roberts (1947) argued, Milton inculcated both the variety as well as the continuity that distinguished the mature madrigal and assimilated them in his poetry. Like the madrigalists, Milton claimed "rhythmical freedom in order to use varied and appropriate rhythms in immediate harmony with the sense, to avoid the limitations set upon expressiveness by straitened patterns" (Roberts 1947: 336).

Blank Verse—"Freedom, But No Mere Looseness"

This specific notion of musicality—organizing rhythm and handling sound after the example of music—generated Milton's treatment of blank verse. According to Saintsbury, Milton completed the work begun by Chaucer and continued by Spenser and Shakespeare (Saintsbury 1923 ii: 356). "Freedom, but no mere looseness" marks his blank verse as best exemplified in *Comus*, the two *Paradises* and *Samson Agonistes*, for with its unrhymed endings it enabled "to variously draw sense from one line to another," as Milton himself has declared.[26] Lacking a stanzaic frame, it also freed large-scale units from confined forms, while guaranteeing rhythmic coherence through the continuous repetition of the five feet of blank verse. The beat, emancipated from the regularities to which it was bound before, contained a variety of sub-divisions, controlling the flow of time without hampering the construction of large scale metrical entities. The flexible use of the pause helped prevent symmetrical "stateliness." The interaction among these various procedures was consummated in the verse paragraph by means of a "continuous rhythmic flow varied in speed and strength of impulse, rising, falling, sometimes hesitating briefly, but never wholly losing its forward movement until the period reaches its end" (Roberts 1947). Examine the following excerpt from *Comus* (textual example 2.7; Milton 1957: 96), with the above points in mind. Note, that here too the poem exemplifies the musical qualities it refers to—those of Circe's and the sirens' rapturous song:

Textual Example 2.7

Can any mortal mixture of Earth's mold

Breath such Divine enchanting ravishment?
Sure, something holy lodges in that breast,
And with these raptures moves the vocal air
To testify his hidd'n residence;
How sweetly did they float upon the wings
Of silence, through the empty-vaulted night,
At every fall smoothing the Raven down
Of darkness till it smil'd: I have oft heard
My mother *Circe* with the Sirens three,
Amidst the flow'ry-kirl'd *Naiades,*
Culling their Potent herbs and baleful drugs,
Who as they sung, would take the prison'd soul.
And lap it in *Elysium; Scylla* wept,
And chid her barking waves into attention,
And fell *Charybdis* murmured soft applause:
Yet they in pleasing slumber lull'd the sense,
And in sweet madness robb'd it of itself,
But such a sacred and home-felt delight,
Such sober certainty of waking bliss,
I never heard till now.

Saintsbury has already noted that the freedom implied by the system of blank verse as employed by Milton, encourages experimentation on the part of a cultivated ear endowed with a high sense of structure. In order to succeed, not only must the poet exercise these specific powers, but the reader must likewise be "tuned." What the reader was required to possess was the ability to distinguish among three dimensions of poetic sound-structure: performance, metrical pattern, and prose rhythm, in order to be able to perceive the complex relationship among them. According to Tsur (1977; 1998), the incompatibility of prose rhythm and meter results in a tension that may be resolved through performance. The "metric pattern and the stress pattern," argues Tsur, are, in this sense, "like the incompatible terms of a metaphor. The reader is first arrested by their incompatibility and then makes the necessary adjustment, adding a performance-pattern in which the two incompatible terms are accommodated... The utmost limit of the acceptability of a line," he continues, "is not determined by its metricalness, but rather by the resources of performance to which a reader or a reading public has recourse; that is by the reader's willingness or ability to render it rhythmically"(Tsur 1977: 27-28). As far as Milton's prosody is concerned, it seems that divergences allowed on one level can always be compensated for on another level, accessible to the competent reader. Tsur's revealing comparison between metrical performance and the understanding of metaphors lends support to our argument. We have already viewed metaphors as a part of artistic attempts to enlist the active participation of the reader in the seventeenth century. Prosody, thus, seems to provide an additional model.

The Making of Worlds

The attempts in the seventeenth century to wed prosody to rhetorical requirements testify, as we have argued, to an adjustment of verse to the subtleties of music and its ability to make sense affective. This however was not the sole road that *poesis* took towards *musica*, with its control over duration and its multiple contrapuntal layers. Eighteenth-century England witnessed the rise of a new genre—the novel—which brought home some of the notions that crystallized elsewhere regarding the relationship between nature and imagination, communication and aesthetic form. Focusing on reality and everyday life, the novel tried to convey immediacy in simple ways. Myth, stylistic conventions, and archetypes seemed, at first, alien to the new genre. Its guiding models were travelogues and other forms of writings—journalistic, autobiographical and epistolary—which carried the stamp of documentaries. The impulse to copy "the prose of the world" was not markedly different from the attempts entertained by Dutch painting. In fact, the mutual influences between England and Holland gave rise to some shared cultural aspects that found expression in different though complementary media.

Contemporaries were aware of the similarity. Richardson, writes Mrs. Barbould, "had the accuracy of finish of a Dutch painter... content to produce effects by the patient labour of minuteness" (Watt 1957: 18). Yet the naturalistic approach did not last long. Like the Dutch painters, the English novelists were quick to learn that a mere copy of reality is tedious if not impossible. They soon centered on "the nature of Man," as Fielding called it, subjecting all details to the view it determined. In the first place, it helped to select major characteristics of an assumed protagonist; random features were thus transformed into an organic whole, granting an assumed individual a lived personality. Psychological considerations determined the description of environments, the rhythm of events and their lawfulness (Van Ghent 1961). If changing scales and modes of observation determined the representations of Dutch painters, it was the fluctuating condition of the human heart that engaged the English novelists. As with Dutch painting, the new mode of selection changed habits of reading. "You must consider the story only as giving occasion to the sentiment," writes Johnson on Richardson's *Clarissa* (Smallwood 1989: 125). Whether or not he intended to criticize the novel for lack of unity, is not clear. What kind of unity, anyway, does befit the new genre? For the novelists this question seemed crucial. Trying to dispense with the popular image of the novel, they considered unity essential for artistic legitimacy. Outwardly, unity was achieved by precisely that which was previously rejected, namely, archetypal schemes: Fielding adopted the outlines of comedy, Richardson turned to myth, and Sterne chose the form of the picaresque. Such adoptions, of course, stressed again the implausibility of the naturalistic approach. Novels belong to art more than to nature, if they appeal to reality it is only in order to subject it to that which granted Western literature a world of its own. Indeed, schemes of all kinds served novelists, as well as their readers, in their attempt to create the desired illusion. Within such frames, however, the novel achieved a real internal integration.

However, it is not the mere selection of properties concomitant with aspects of given reality, or existing types, that distinguished the new genre. What distinguished the novel and brings it close to Dutch painting, is its double level of representation—representing that, which is in and of itself, a representation. "Represented seeing and represented thought and speech," writes Alpers, "manifest extraordinary attentiveness, without however acknowledging that interplay between sender and receiver... that normally characterizes the Italian mode (of painting) or language when spoken" (Alpers 1983: 53). Such "duplicity" arises when "sense and sentiment" are interrelated in an indeterminate way, which renders the described events suspect. Sentiment, to be sure, plays a significant role in the drama as well, only that in the novel the indeterminacy of sentiment is interrelated with the intricacies of the "point of view," implying an "inside outsider" presenting and re-presenting the course of events, disclosing only gradually their implications and meaning. The "natural" condition of the novel is in the interrelatedness of its temporal and rhetorical dimensions. The narrative mode, hence, gives rise to the interplay of parameters such as "order," "frequency" (of occurrences of events) and kinds of "durations," to use Gerard Genette terminology, all related to a variety of ways in which the story is presented and represented, highlighting and adumbrating characters, features and happenings.[27] Such compositional possibilities are the *sine qua non* of the novel. The outstanding development of the novel from Defoe to Sterne testifies to its potential, for in less than fifty years most of the techniques that eventually became identified as novelistic, reached fruition.[28]

By the eighteenth century, it was understood that the second reading of a novel is essentially different from the first. We are assured by present-day theorists that the second reading is, indeed, an altogether new experience, consisting of a "counterpoint" between sober and naive readings. Richardson was among the first to realize this property, and with Austen's *Emma*, critics tell us (e.g., Wiesenfarth 1975), it reached full fruition, in that almost every sentence has its definite reference to equally unemphasized points before and after in the development of the plot.

The existence of a second level of interpretation, is, of course, neither unique to the novel, nor to a certain style in painting. In fact it is basic to art in general, as Adam Smith will tell us. Yet Dutch painting and the novel brought to the fore that which was previously hidden in other forms of art. Paradoxically, new roles were given to a reader even if he was less educated than before; the reader, regardless of erudition, was summoned to enlist his interpretative powers in an unprecedented way. Relying on his commonsense, he had to recreate the work's world, fragmented as it was by different temporal modes and structure. Interpretation became thus a process, achieved through negations and substitutions, guided by an internal range of probabilities, as James Beattie insists.[29]

Novelists were aware of the lack of appropriate models for interpretation and tried to supply the reader with guidelines of all kinds. Fielding, for example, interpolated interludes, imparting his own interpretative theory into his works, not hesitating to intrude in the plots whenever it seemed useful.[30] With Sterne, whose narrator is also the main protagonist, such intrusions led to new confusions, requiring their own solutions. In fact, the awareness to the "difficulties" of the novel was part of the realization that the genre presupposed new habits of reading and an altogether new mode of communication. In that sense, novelists did not only create fictional worlds, but formulated an artistic domain with its own grammar and conventions. This was not, however, the private affair of writers, the reader had to acquaint himself with the limits and the potential of the medium, rendering his part in the transaction. Artistic secrets of the metier, it seems, were more easily detected in the novel than they were in other art forms and it is precisely this saliency which required clarification.

Readers and writers from Beattie (1778) to Barthes (1977) and Bakhtin (1983) increasingly became aware that narrativity does not present a simple, ordered linearity but exploits a multi-layered hierarchical texture, which gains in coherence in the process of its weaving. Moreover, in such a process what seems temporarily closed, opens up to enable larger units. These, in turn, resolve previous tensions only to create new ambiguities, which eventually lead to the final denouement. It is thus evident why such theories should rely so heavily on the literacy and the good will of an active consumer, in whose consciousness resides the totality of that which the initiating author aimed to achieve.[31]

That music should provide the prime paradigm for such a conception of unfolding is not surprising; explaining its coherence using extra musical factors was always dubious. However, in music too these matters had to be clarified and only surfaced gradually in the steady attempt to create meaningful musical constructs. It is these self-referential musical constructs which theories of narratology took into account. The following chapter will deal with these developments in music. It is no exaggeration to claim that it is the British writers who were the first to see the developments in both literature and in music as related to the same interrelated cognitive phenomena.

Notes

1. Ackerman 1961: 76. For a rich array of Caravaggio's and his contemporaries genre paintings see *The Age of Caravaggio,* annotated catalogue by the Metropolitan Museum of Art, 1985.
2. Cropper and Dempsy 1996: 177-249; Chapman 1990. Poussin also tried to base expression on theoretical constructs as well. Searching for musical modes in painting (Ross 1984), he was still influenced by the concept of Platonic ideas, though he tried to embed them in perceptual considerations. By and large, his theoretical orientation like that of the eventual French academy, depended on rational deduction of rules of art (Lee 1967).
3. Painting, Pater argued, is "musical" whenever the subject is subordinated to a pictorial design (Pater 1980: 111). "Musicality," on the other hand, depends on a "perfect identification of matter and form." This is not the place to explain how, from a pictorial point of view, the two kinds of musicalization relate to each other and how they became mixed within the same style. What is important for our discussion is that the first conception, like the second, is also bound to the tradition of "world harmony," that is, to the idea of harmonizing diverse elements into a proportionate and ordered structure.

4. The choice of the ideal moment in a story, pertinent for rendering it in a picture, was one of the most important tenets of theories attached to the *ut pictura poesis* dictum (Lee 1967) as it was later reiterated by Lessing's *Laocoon,* though relying on opposite aesthetic premises.
5. Panofsky's article on Poussin's *Et in Arcadia Ego* (1983) is the most classical treatment of the subject. Bryson (1981), dealing with *Word & Image* in French Painting of the Old Regime highlighted the phenomenon as part of the new challenges such combinations put forward. Alpers (1983) deals with the representation of texts in Dutch art.
6. Foucault viewed Velázquez's famous *Las Meninas* as paradigmatic to what he viewed as a "new order of things" imposed by seventeenth-century *episteme,* which viewed the dichotomy between signs and their signifed as given and final. For Dou's interesting painting *The Quack* (1652) consult Alpers 1983: 116. Also see our caption to (Plate 1).
7. Hamilton 1963: 97. Even Milton shared this conception, according to Hamilton (1963: 93-4), since he too was steeped in the conception of poetry as a rhetorical art, "clothing thoughts and feelings in language that is correct, appropriate and pleasing," always keeping in mind its didactic, moral purposes.
8. The way poetry directly partook in the intellectual climate of the beginning of the seventeenth century is lengthily discussed in Nicolson 1960, Grant 1985.
9. See Donne 1952: 191, a grand elegy written for the first anniversary of a Mistress Elizabeth Drury's death. "The Firmament" writes Donne in the following lines "...is crumbled out againe to his Atomies./'Tis all in peeces, all cohaerence gone." The lost coherence, we are told in the next lines, does not only refer to astronomy but also to geography, politics, beauty; the world as a whole "is rotten at the heart." Only song, Donne concludes, is able to overcome the decay this new state of affairs (new philosophy and untimely death) brought about.
10. Cassirer emphasizes the overlappings and interactions of symbolic forms, while crystallizing into autonomous systems. He considers the interaction between science and language paradigmatic in this respect (Cassirer 1955 i).
11. The widespread of rhetoric in England at the time, and among poets in particular is evident from studies such as Butler's (1980) and Vicker's (1970).
12. Concerning the cognitive treatment of metaphor, Ricoeur (1986) points to its nineteenth-century French antecedents. As is well known, the treatment of metaphor from a cognitive point of view, received much attention since Black, both from a philosophical as well as a psychological point of view.
13. Goodman is of the same opinion, though he anchors it in a broader theory of reference: "...metaphor typically involves a change not merely of range but also of realm. A label along with others constituting a scheme is in effect detached from the home realm of that schema and applied for the sorting and organizing of an alien realm" (Goodman 1968: 81-5). Black has also been criticized for his "system of associated commonplaces," which emphasizes the interaction between marginal elements that remain within the boundaries of known connotations. Black somewhat modified his position in a later article. The criticism, however, does not radically change Black's initial insight, which stresses, in line with Aristotle, the affective power of metaphors on the hearer, by breaking his set of expectations.
14. Though a rich conceptual web existed by which resemblance was treated, the above mentioned four terms, according to Foucault, are the principle figures that "determine the knowledge of resemblance with their articulation" (Foucault 1972: 17).

15. Our discussion of the twilight of the *ut pictura* tradition in poetry rests primarily on Hagstrum's exemplary work (1958) on the subject, as well as on Lee's classical study (1967), Barry's (1987) and Heffernen's (1985) profound insights.
16. Visual aspects of an immediate appeal were equally central to the English masque, though the masque aspired towards a synaesthetic experience as well, a tendency not alien to seventeenth-century art.
17. Neoclassical pictorialism—mainly represented by poets like Pope and Thomson—concentrated primarily on allegorical renditions and on satiric constructions. In allegories—as realized in Thomson's *Seasons*—a pictorial construction, itself an allegory, seems to be consciously and fully translated into the verbal medium. Such translations center primarily on the major figure to which all poetical details are subjected. Pictorialized through and through, the poetical figure returns, however, to its origins. It lives in its visual splendor, scenic setting and frozen movement; a pure, idealized present oblivious to past and future events. The models for the "picturesque" at the beginning of the eighteenth century were provided also by real gardens, for example, those planned in the spirit of what was defined to begin with as picturesque in painting and poetry. Thus a triangular if not a circular relations linked the three arts (i.e., poetry, painting, and gardening) as Heffernan (1985) put it. Only towards the end of the century, with poets like Wordsworth and Coleridge, and painters like Turner and Constable, doubt was cast on the immediacy of such a link, making room for more intricate relationships between the three arts as Heffernan argues. Heffernan's claim supports, and is partially explained, by our argument about the turn in aesthetic deliberations concerning art in Britain in the second half of the century.
18. Jacob's treatise is discussed at length in chapter 5 and in the second volume of our study. It is an important harbinger of aesthetic theories related to the dicta of artistic media.
19. Saintsbury 1923 vol. ii: 134. Though Sainstbury's study may be considered obsolete by some, there is no current study of prosody in English poetry which approaches its scope and comprehension. We found many of his observations relevant and insightful.
20. Mace (1952) was indeed right in viewing Daniel's arguments as supportive of the relegation of music to the intellectual activities that took place in seventeenth-century England.
21. While English poets were freeing themselves of musical accompaniment, English musicians, especially those who practiced the virginal, were experimenting with pure musical unfolding. Their products contributed greatly to the development of instrumental music.
22. Donne secured two effects, according to Grierson (1965), firstly, the troubling of the regular fall of the verse stresses by the intrusion of rhetorical stress on syllables which the metrical pattern leaves unstressed, and secondly, an echoing and re-echoing of similar sounds parallel to his fondness for resemblances in thoughts and things, apparently the most remote from one another. The interaction in Donne's poetry between the musical and the rhetorical thus comprises two levels—it occurs on both the units of the line and the units that are above the line. Both techniques end in strengthened rhetoric, creating mood, directing the hearer and clarifying meaning.
23. Cox supports Grierson's theory and claims that what was achieved by Donne and his followers is providing "the short lyric something of the flexibility, the urgent and profound expressiveness that came to be developed in dramatic blank verse" (Cox 1956: 49).

24. Milton's verses, writes Saintsbury, "do not float, they march, and march magnificently, but always with more touch of mechanism in them than we find in Shakespeare, with more of the earth and less of the wind and water, if with hardly less of the fire, in their composition" (Saintsbury 1923 ii: 234).
25. It is to Kramer's (1984: 6) credit that he employs Polany's (1966) conceptual scheme in this connection, though it does not go much beyond Hanslick's observation regarding the respective primacy of sound vs. meaning in music and language (Katz and Dahlhaus i 1986: 390).
26. His early poetry notwithstanding, Milton shared Campion's view regarding the "triviality of rhyme" (Saintsbury 1923 vol. ii: 236). He however applied this view in his poetry to a greater extent.
27. Discussing order, Genette (1980) writes that the "return retroactivity confers on the past episode a meaning that in its own time it did not yet have." It enables "the modification of the meaning of past occurrences after the event, either by making significant what was not so originally or by refuting a first interpretation and replacing it with a new one." The other parameters (frequency, duration, voice etc.), according to Genette, stand for no less intricate activation of levels of meaning.
28. In both Fielding and Sterne one finds all the intricacies of 'order' (which Genette attributes to Proust's sophisticated narrative technique): 'anachronies' and 'gaps', "recalling to the narrative's own past" (Genette 1980: 54), as well as "retrospective refutation of a mistaken anticipation" (81). In the rhetorical dimension, this means a subordination of one narrative to another in a sort of narrative syntax (48). As far as 'duration' is concerned, Richardson exploited "pause," Fielding, "summary" and "scene." Richardson's epistolary novel reveals also the secrets of repeating the same event through different eyes and so does Sterne's "stream of consciousness" novel, which employs repetition of the kind Genette would have called a "mental construction" (113).
29. See vol. II: 194. Probability brought along new kinds of signs, judgments and expectations, a "sensibility" effected, as Patey (1984) argues, by developments in scientific theories. For Hunter (1975) the stress on probabilities testifies to a new kind of hero, which addresses "the art of the possible rather than some abstract notion of absolute ideals."
30. In *Tom Jones*, writes Alter (1967) "the narrative is everywhere...steadily editorializing, sometimes conspicuously, sometimes unobtrusively." See also Crane 1967: 86-7.
31. It is interesting that Sartre (1950), from his existentialistic perspective, should grant the reader the will to enlist his mental efforts in the final realization of the literary work of art. Sartre emphasizes that unlike nature, the literary work of art leaves its entire instantiation to the consciousness of those who exert the mental efforts, which they are called upon to activate by the work itself. They are rewarded, in turn, by an object meaningful through and through and within their possession.

3

Music's Begotten Illusion

Musica Ascending the Stage

The development of music in the seventeenth and eighteenth centuries was more homogenous and comprehensive than that of its sister arts. In music, a single paradigm—the one associated with the birth of opera—came to dominate the whole scene. It is the growing awareness, in theory and practice, of generic specifications and stylistic differences which, paradoxically, testifies to that predominance; while distinctions first served to clarify the major goal, they later reflected its realization. *Seconda prattica*, thus, stood for an overall stylistic alternative, whereas terms such as *stile concertato*, *stile rappresentativo* and the like circumscribed its different aspects.[1] The later debates concerning the superiority of national styles, on the other hand, centered on questions of *preference*. French protests notwithstanding, it was the Italian spirit related to expressivity that was adopted by all, not only on the stage but in the church and chambers as well.[2]

The Italian conception that came to pervade almost all of baroque music was concerned with affect. The purpose of music was now to move human affections and composers everywhere enlisted their creative powers towards the realization of this professed purpose. On this very ground, we are told by Schrade, "modern times which start with the Baroque, and old music, which ends with Palestrina, part company forever" (Schrade 1964: 99). To understand the separation, one may turn either to musical characteristics, mainly tonal harmony, or to sociological ones, such as the rise of a new audience. Both characteristics are important, but do not capture the nature of the change, for the first was its consequence while the second provided the appropriate circumstances. Likewise, "expressivity," broadly speaking, is of no great help, for music since Plato, through Augustine and Tinctoris, was invariably related to dispositions. The nature of the change is anchored elsewhere, between the stylistic (which is too particular) and the cultural (which is too vague); it is to be sought in the symbolic realm. It itself rested on the change in music's ontological status and on new semiotic modes, which substantiated and further projected this change. This change was so fundamental that it affected the whole destiny of the art, reverberating in wider cultural circles.

As a representation or an exemplification, music before 1600, philosophically speaking, related primarily to ideal structures—"numerical proportions"—rather than to expressive and dramatic contents. Since static musical attributes exemplified the divine order, dynamic "earthly" musical elements were mainly conceived metonymically: Music, created for religious purposes, for example, turned into an extended prayer; the individual was called upon to express his belief via musical sounds either through actual church participation or through the identification with a group chosen for that purpose. Expressing emotions in early religious music was thus void of fictional elements, the personal prayer was simply substituted by more successful renditions. The status of expression in secular music was not essentially different. The troubadours and the trouveres, for example, passed as *real* lovers, uttering heartfelt feelings to their beloved, soliciting their response. The lady, though often assuming an ideal existence, partook in reality—not unlike the holy virgin; song, *qua* song, was addressed to her. Though Petrarchian elements were gradually pervading the musical scene by the fourteenth century, it was the poetry rather than its accompanying music that assumed fictional status. Even when the poetry was expressive, rather than descriptive, the singer was only promoted to the status of a *fictive* bard for whom song is a native tongue. The frottola and the Renaissance air were conceived along these lines, reenforcing "bardic fiction" through an accompanying instrument and an improvised style (Palisca 1960). With the madrigal, the assumed bard was extended to a group of singers, alternating between description and expression, without necessarily calling into doubt the "literalness" of singing. The whole attitude is best illustrated in the use of music in the *intermedii* and similar dramatic forms: Song was incorporated into the reality of the drama like Cherubino's "Canzonetta" or Carmen's "Habanera" in the later opera. In all of these genres—frottola, madrigal and *intermedii*—illusion was relegated to the very act of composing; the listener was called upon to "believe" that it is the singer who "composed" the song, words and music, either on the spur of the moment or at an earlier time.

Treating Music as "Normal Speech"

This seems to have changed once Monteverdi, in his *Orfeo*, called upon *Musica* to ascend the stage.[3] Though still centering the drama upon a piece of "real" music (Orfeo's address to the *possente spirito* on the gates of Hades), Monteverdi expected major parts of the music to assume a fictive status.[4] Some utterances of the drama required, in fact, a "willing suspension of disbelief" in their musical nature, so that "from the point of view of the implied or enacted drama," as Cone puts it, "the dramatic character was not 'really' singing at all but engaged, as it were, in normal speech" (Cone 1974: 30). If *Orfeo* was somewhat vague in this respect, Monteverdi's following works, those that do not include the "powers of music" as part of the plot, determined the future of opera: *Il Combattimento di Tancredi e Clorinda* (The Battle between Tancred and Clorinda) (1624) his *Il Ritorno d'Ulisse in patria* (The Return of Ulysses to the Homeland) (1640), and his *L'incoronazione di Poppea* (The Coronation of Poppea) (1642) clearly departed from the bardic convention, and much less needed the insertion of legitimate music for dramatic appeal. Musical utterances were now conceived as pertaining to fictive characters possessing fictive emotions musically expressed. It is as if music shed its musical appearance; one was called upon to ignore exact pitches, organized rhythms and the like in order to be able to concentrate, instead, on the illusion begotten through them. With music turning into fiction, the expression it conveyed turned into metaphor, the principle mode through which the dramatic characters revealed their passions. In that sense, music became central to the drama more than it had ever been before. Paradoxically, attention was now drawn precisely to that which one was supposed to ignore—a pure musical expression, but one that was anchored in a new symbolic realm.[5] Smith will later explain that this paradox holds true for art in general, and underlies the aesthetic experience.[6]

The change seems to have derived from the agenda inaugurated by the members of the *Camerata*, though their own attempts still rested, outwardly, on old conceptions. Vincenzo Galilei's adoption of the Renaissance *arie* for the purposes of the new expressive ideals do not stray far from the "bardic" convention (Palisca 1960). The same holds true for the monodies in general. The concentration on the Orfeo myth—including Monteverdi's *Orfeo*—is based, likewise, on viewing music as literally expressive. The belief in a Golden Past when music exerted its effects in the actual world testifies as well to a misunderstanding concerning the nature of symbolism. The hermetic tradition, which also considered music for what it, supposedly, is, rather than for what it represents, did not contribute to the formation of new insights either. "Until symbolic forms are consciously abstracted," we learn from Suzanne Langer "they are regularly confused with the things they symbolize. This is the same principle," she adds, "that causes myths to be believed, and names denoting powers to be endowed with power, and sacraments to be taken for efficacious acts" (Langer 1976: 245). In the *Birth of Tragedy,* Nietzsche, in his ingenious way, understood that this confusion accompanied the early opera whose founders, he argued, shared the "idyllic belief that every sentient men is an artist." The creators of the *stile rappresentativo*, according to Nietzsche, conjured up the "artistic primitive man," that is, the man "who sings and recites verse under the influence of passion." They dreamt, he adds, of a time "when passion sufficed to generate songs and poems; as if emotion had ever been able to create anything artistic" (Katz and Dahlhaus 1987 i: 178). Indeed, the founders of opera did not understand the nature of illusion since they considered their "artificial flowers"—to use Smith's example—as real flowers, though of a rare nature and of a kind long forgotten.

Accompaniment as Symbolic Frame for the Illusive

Misconceptions notwithstanding, the illusive and the symbolic came to the fore nonetheless, albeit in a rudimentary way, in the *stile rappresentativo* where the solo voice guaranteed the former and the accompanying continuo—the latter. Galilei's insistence on the solo voice as the only "instrument" for singing represents the conscious adoption of the illusive component: only the realistic expression of one's own emotions is able "to induce in another the same passion that one feels oneself," argued Galilei (Strunk 1981 ii: 127). Interestingly, he borrowed his main technique from another artistic medium, the drama, rather than from everyday practice. He advised the composer to further the dramatic delivery musically: to pitch exactly the "high and low of the voice," to organize the succession of utterances rhythmically. Still, it was primarily the realistic delivery that governed his treatment of the solo voice.

The most important implication of reverting to a speech model—be it rhetoric or drama—was that it rendered singing an illusive standing. From a realistic point of view, singing, at least on stage, was to be perceived as heightened speech, not as a musical entity. As a consequence of the new correlation between the solo voice and the dramatic delivery, the instrumental accompaniment lost the standing it had within a "bardic" frame and called for a new symbolic justification. In retrospect, it seems that the new *basso continuo* supplied the needed justification. Indeed, the crystallizing conventions of the thorough bass aimed at artistic goals entirely different from those of the accompaniment of improvised Renaissance singing.

* * *

Musical Excursion 1: From Galilei's Aria to Monteverdi's Lament

In his *Dubbi intorno a quanto io ho detto dell'uso dell'enharmonio con la solutione di essi* (Doubts about what I have said regarding the use of enharmony with their solutions) (1588), Galilei explores the technique of accompanying popular *arie*, in light of what he conceived as the new humanistic ideals: music associated with shepherds and workers in the fields after having finished their labors; odes and hymns sung to the cithara by ancient poets and so on. While emphasizing that limitation of means was the secret of the Greek art, contemporary popular *arie*, found in collections of laudi, villote, villanelle, and the like, constituted his real models. The upper melody is the determining factor in compositions of this sort, and not the bass.

Commenting on Plato's *Laws,* Galilei differentiates between two kinds of accompaniment, one that he recommends, and the other that he discredits. Explicating the first, he says: "And this occurs when a soloist sings to an instrument in which are struck at the same time several strings disposed in such a way that they make various consonances among themselves, but in a manner uniform with the tones of the soloist, with which they thus become all of a single and same sound" (Palisca 1960: 356). In this particular accompanying technique he found a model of simplicity, naturalness, economy and direct expression of text and mood.

His example (musical example 3.1a) illustrates this union, that is, the subordination of bass to the harmonic implications of the melody. His example of the second kind of accompaniment (3.1b) shows that "the body of divers consonances that the instrument causes is not perfectly adapted and united with the air of the singer" (ibid). This example introduces two chords of the sixth and twice the E flat harmony, which is not in the natural scale of the melody, as Palisca writes. However, this technique takes note of the dramatic monody which subjects the bass to the intricacies of affectionate expression, as exemplified in the famous *tu se' morta* (musical example 3.2), Orfeo's lament of Euridice death. In this *lamento*, expression is achieved by forgoing both tonal unity as well as "harmonic" complacency. From the vantagepoint of tonal harmony, such tonal shifts are easily explained. But in Monteverdi's time this must have struck the listener as a novelty, despite the chromatic boldness of the Italian madrigals of the high Renaissance (see musical examples 2.2 and 3.3). The mutual determination of voice and bass, as a unity, embodies the protagonist's affective predicament: The bass defines the basic tonality, disclosing the emotional state of the protagonist before the latter even utters a sound, and when the latter's voice is heard, it immediately clashes with this fundamental emotional stratum via a harsh dissonance. The seeming gesture of accepting fortune's verdict ("*tu se' morta*"—your are dead) also reveals that the hero is far from being reconciled. The dominant chord, which appears as a separate utterance of the accompaniment (m.3), thus stands for the inner urge of the protagonist who is trying to express what is still inexpressible, that is, the terrible cry, the unresolved dissonance: "*ed io respiro*"—I am breathing while you are dead! Once uttered, Orfeo is led to further explore his emotional state; he repeats the same idea—"*tu se' da mi partita*" (thou art from me departed)—to almost the same music, intensifying its expressive characteristics. The further shift into C major appears as an inevitable outcome of his defying declaration "*ed io rimango, no*!" The voice now leads the bass; the "outer expression" becomes clearer and stronger and then the "inner countenance" ends in a bold proclamation "*n'andrò sicuro a più profondi abissi*" (then I shall seek the most profound abysses). The bass, or what it represents, acts in m. 12 as a mitigating force, chromatically leading (via an augmented sixth) to a new composure on d minor, representing the "distance" from the starting point (g minor) of the whole emotional excursion.

* * *

The improvised *arie* were conceived primarily to suit a given poetic structure, characterized by uniform rhythmic patterns and melodic repetition. As a result, it often rested on a ground bass, a repetitive formula, enabling improvisation on the singer's part. As far as the actual composition was concerned, it was the bass, in such cases, which determined the melodic line. Though related to poetic structure, the accompaniment was not tailored to the meaning of the text; the same bass, in fact, could be used for different and even contrasting poetical contents. Even when the melodic line preceded the bass—as was the case in the frottola and related forms—the latter still consisted of a single line which was harmonized instrumentally or vocally.

In contrast to the "amplification of a single line," the accompanied monody, says Pirrotta, was a "reduction of polyphony"; in the former, in other words, polyphony was the end result, whereas in the latter, it was a starting point (Pirrotta and Povoledo 1982). The madrigal lies between the two procedures. The simple harmonies of the frottola were elaborated in the madrigal contrapuntally as well as harmonically. Via such elaboration, composers aimed to communicate meaning; different kinds of imitation served as either descriptive or dramatic renderings, while homophonic settings were often relegated to the more declarative parts of the text. The choice of texture, thick or thin, also served to interpret meaning (see musical example 2.2, Marenzio). The multiplicity of voices, nonetheless, served to achieve a unified effect.

A most significant aspect of the elaboration of the madrigal related to harmony. The madrigalists developed a "secret code" which

Musical Example 3.1a

"Unisono"

Musical Example 3.1b

"Consonanza"

Musical Example 3.2

Claudio Monteverdi: "Tu sei morta"

Note: The value of the quarter note remains constant in bars of varying lengths.

Musical Example 3.2 continued

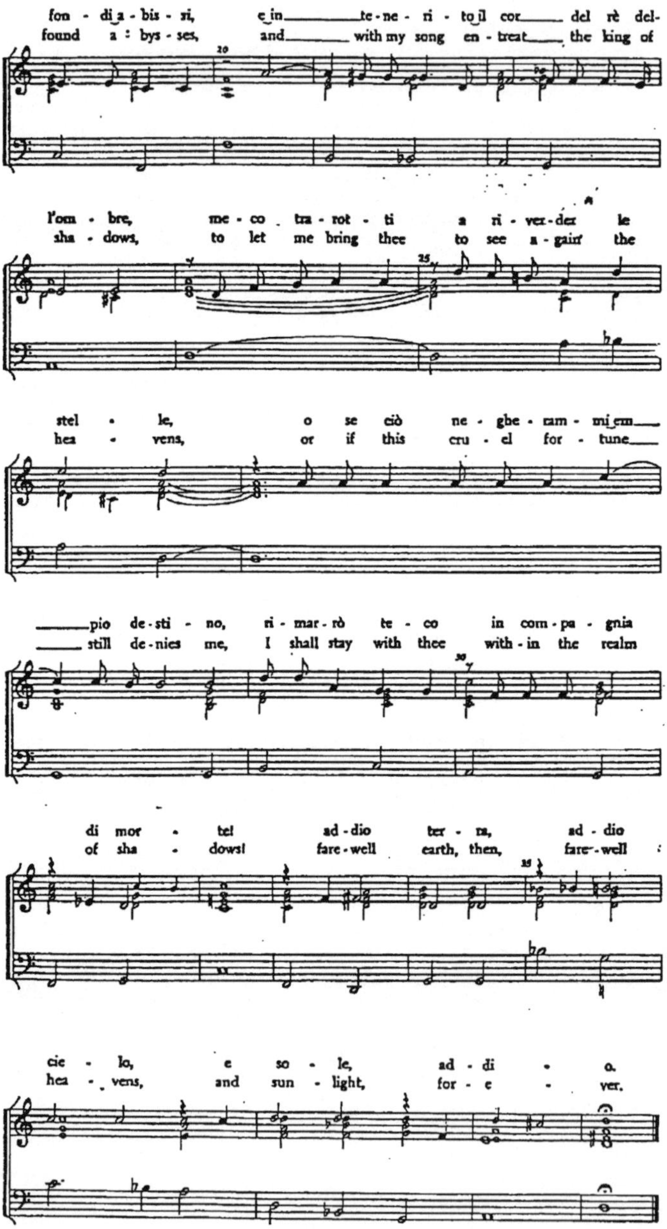

was related to distinct chords, connected to each other by diatonic, chromatic or enharmonic voice leading. All chords could serve as points of "departure" or "arrival"; the way they were to be used and the kinds of connections to be employed were determined by the semantics of the text. Thus a tonic "arrival" was relegated to the end of an utterance and served as a point of departure for the next one. Diatonic relations were used for smooth connections of simple utterances, whereas the chromatic and enharmonic ones were reserved for the dramatic or expressive parts of the text. The whole procedure necessitated, of course, a frame of reference to which all arrivals and departures could relate, a governing tonality (see musical example 3.3).

* * *

Musical Excursion No. 2: The Lady that was Left Alone

Cipriano da-Rore's "Da le belle contrade d'oriente" (from the beautiful eastern horizons) (1566), which is generally relegated to the "middle" period of the Italian madrigal, employs chromaticism in a bold way: The dominating tonality, F major, gives way (in m. 21) to g minor (highlighting the words "when I have heard an ardent sigh"), while clinging, in the cadence (in m. 30), to its major subdominant (C). From m. 32 the D minor is chromatically hinted, leading to an A minor of sorts, ("Adio, Che sarà..") returning through further chromatic steps back to A as a dominant of d (m. 40). This harmonic progression underlies the most poignant utterance of the whole madrigal—the lover's question about having been left doleful. She herself, after an "abrupt" stop, answers the question, in m. 41. For this rhetorical change Rore uses not only a rhetorical *abruptio* (i.e., a rest of half a measure) but also a piercing chromatic shift, from the half cadence seventh chord on A to the distant c minor, which then subduingly leads back to the more relaxed subdominant area of F minor.

Of course this is not the only technique which is used in this madrigal for expressive and dramatic purposes. No less important is the way by which the polyphonic texture is arranged, creating both metaphorical as well as literary representations of the text.

* * *

Musical Example 3.3

Cipriano Rore: "Da le belle contrade d'oriente"

From: Cipriano Rore, *Opera Omnia*, ed. B. Meyer, Vol. 5, American Institute of Musicology 1971, pp. 96-99.

Musical Example 3.3 continued

Musical Example 3.3 continued

In fact, it was the harmonic flexibility that enabled further developments within the framework of the *stile rappresentativo*. Following Doni, Palisca mentions three kinds of recitative that were employed at the beginning of the seventeenth century: narrative, recitational and expressive (Palisca 1981: 34). These different kinds, we believe, did not constitute distinct genres, but were part of one and the same "system." Thus, for example, a single tonality was largely adhered to when a particular atmosphere or passion was at stake and less so in the case of quick changes of emotions. Eventually, this brought to the final separation between aria and recitative, while still sharing the same decodable harmonic "semantics."

Instrumental Accompaniment as a Psychological Substratum

Elaborating upon their own heritage, the monodists helped to define the symbolic significance of the instrumental accompaniment. They divided the previous multiplicity of voices into two virtually contrasting lines, making clear that the upper line, that is, the voice, derives its existence from the lower and that the latter determines the expressive content of the former. Given an altogether new function, each move in the bass came to reflect not only professed rhetorical changes but their implied psychological base as well. It is here that the "subconscious" in music was born. It should not be confused with the "unconsciousness" related to the *singing* of the protagonists; there, as Cone made it clear, it is mainly analogous to the unawareness, on the part of an excited speaker, "to his inflection of tone, pitch, loudness rhythm and speed" (Cone 1974: 33). The instrumental accompaniment was to reveal to the hearer something in relationship to which the explicit articulation of the protagonist stood as the tip of an iceberg. It was supposed to help gain insights into "the entire realm of attitudes, feelings, impulses and motivations of the protagonist, unverbalized because essentially unverbalizable, or because they have not risen to the level of explicit articulation in the mind of the subject" (Cone 1974: 33). It is this special combination of voice and accompaniment that enabled music to shape the "energies which animate our psychic life." Interestingly, the making of the symbol preceded almost by two centuries the conscious articulation (by people like Schopenhauer and Wagner) of that which it came to symbolize. But this is characteristic of the making of symbols, as Vico and others tell us.

From a symbolic point of view, the penetration into the hero's psychic life is impossible without assuming a point of view, that of the composer, what Cone calls "the composer's voice." Like the narrator in the novel, the composer's persona, argues Cone, is responsible for everything in the representation; it is his point of view that determines the work. The music, however, that "proceeds entirely and directly from the consciousness of the composer's persona, could be dramatically appropriate and highly expressive," Cone writes, "yet outside the represented world of the character and thus totally unperceived by him" (Cone 1974: 32).

The greater preoccupation, in recent years, with narratology, its role and function, and with the interpretative roles of the narrator, reader, and listener, has given rise to novel approaches to Cone's proposition regarding the composer's voice, which indirectly supports our argument. Cone's concept of the "musical persona," accordingly, has been viewed as a rhetorical strategy, which aimed to "humanize the impersonal agency that we hear in music" (Kramer 1995: 119), and the "virtual voice" which he addressed is no longer considered to be that of a single originating speaker (Abbate 1991). Music cannot narrate in its own right, argues Kramer, since it stands to narrative as ciritique, performative, or supplement. Music, he suggests, needs to be heard as a cross between rhetoric and history: "when Cone presents his voice as the immanence of the composer's authoritative, monological intelligence in the music, what he fails to recognize is that his supposed immanence is both a rhetorical effect—sometimes urged by the music, sometimes by the protocols of listening—and a historical formation that seeks to exclude other voices: the voices, precisely, of the other." To hear the "unsung voices," Abbate talks about, argues Kramer, requires "listening through" that voice which has acted historically as the agent of exclusion (Kramer 1995: 121).

These important qualifications give further impetus to the investigation of the genealogy of the role of voice in Western music and its various authorial strategies. This necessitates, however, the inclusion of thorough investigations of specific social and political contexts, which are beyond the scope of the present book. As far as our argument is concerned, the elaboration of the concepts of author and authority in political thought and other discourses, at the beginning of the seventeenth century carried relevance for art as well. Hobbes's definition of "person" and "author," for example, already betrays awareness with regard to the artificiality of their applications:

> *A person, in whose words or actions are considered, either as his own, or as representing the words or actions of an other man, or of any other thing to whom they are attributed, whether Truly or by Fiction [...]* [w]hen they are considered as his owne, then is he called a *Naturall Person*: And when they are considered as representing the words and actions of an other, then is he a *Feigned or Artificiall person*. (Hobbes [1651] 1968: 217; original italics)

Hobbes was perhaps ahead of his time in his explication of the ways by which artificial authors gain authority, yet he was reacting to their new embodiments not only in politics, but also in the arts and the sciences. In any case, he acknowledged the tension between the represented and the representation, which in Dutch painting implied an observer whose point of view determined the work of art, and in literature a narrator presenting and representing the course of events (see chapter 2). The tension inherent in such representations raises a claim, however, that seems to contradict Kramer's and Abbate's above contentions. The intervals between the represented, the representation, and the representing agency harbor emancipatory powers, which may move against the artistic hegemony that enacted them in the first place, giving vent to repressed voices. In any event, such emancipatory powers are conditioned by the semiotic potential inherent in the representation, and depend on the cognitive apparatus that enables them.

This may partially explain the line leading from Monteverdi to Wagner, as Donington suggested;[7] yet it embraces not only opera but other kinds of music as well. Already in the course of the seventeenth century, its characteristics came to pervade the entire musical scene, the vocal as well as the instrumental domain. When Frescobaldi, in the preface to his *Fiori musicali* in 1636, urges the performer to "find out the affection of the passage" before playing it (Bukofzer 1948: 48), he takes for granted the existence of a fictive persona to whom the affection is related. The eventual affinity between concerto and aria, between the symphony and comic opera rests on the same assumption. Cone's analysis reveals how symphonic works fostered this kind of symbolism though he ascribed it to a later period. Much of that which the new way initially implied called for further elaboration; it is only in retrospect that one is able to ascribe the "invention" of this symbolic construction to the crucial period around 1600. Further elaborations notwithstanding, we insist that the paradigmatic change, including its various implications, already occurred at the beginning of the seventeenth century. As in painting and literature, the musical listener was called upon to enlist his imagination in an entirely new way. In music, however, the change was much more comprehensive. Illusion, fiction, expression, and an anchored point of view became significant in music simultaneously and determined its very nature. This was constructed so that the symbolic setting directed the imagination while supplying the rules for its activation, that is, the frame included the making of that to which it refers to. We have noticed a similar situation in poetry when poets gained awareness of the power of metaphor. Whereas their new realization amounted to better intuition with regard to that which poets always did, the new awareness in music resulted in a revolutionary change. It is this revolution, we believe, that forever separated music before and after Palestrina, to quote Schrade again.

No one, however, seems to conceive the nature of the change at the time. In fact, contemporary theories, as we saw, totally misconceived what has taken place. This state of affairs continued for a long time, making futile all attempts to interpret the specific contents the new medium brought about—the musical coinage of emotions. In other words, without a clear view of the symbolic nature of the art, all musical expressions of emotions assumed a literal standing. The rhetorical analogy, for example, that assisted in forming the new illusion was misinterpreted on a theoretical level. Parallel to speech, music rhetoricians considered music analytically, interpreting each utterance as a separate item. Given the nature of music's articulation, this conception was unwarranted, even in the recitative, not to mention the aria, for it is the changes—transformations, transitions and the like (though not necessarily comprising a single affect)—which create a unified musical effect. In chapter 5, assisted by the British writers, we shall try to show how the new symbolic frame could, in fact, only lead to semantics of synthetic constructions.

Creating Expressive Coinage

Whether music can convey specific meanings is one of the most controversial questions in its aesthetics. The debate started long ago, already in ancient times, yet, as we have seen, it received a new impetus in the seventeenth century with people like Descartes and Mersenne and has never left the stage since then. Further developments in music and in philosophical thought provided new insights and arguments. By the beginning of the eighteenth century, there was no one who could ignore expressivity in music; in fact it was taken for granted. Questions, which were now dealt with, concerned the *relationship* of music to passions: Does music express or represent the passions? Does emotionality relate to music naturally or is it ascribed to music by convention? If ascribed by convention, how does it succeed in affecting the feelings? And finally, are the emotional contents of music of an analytic or a synthetic kind?

Of course, even in the eighteenth century these questions were not formulated in this way. Yet one cannot understand the entire controversy without assuming their tacit presence in the minds of the contenders. In the twentieth century, musicologists and aestheticians, assisted by historical evidence and hindsight, were able to attack these questions more directly. Controversies notwithstanding, it is agreed that the era between 1600-1900 was crucial for the development and use of expressive contents. Even Cooke (1959), who extends the epoch back into the fourteenth century, draws his most persuasive examples from music composed after 1600. Awareness of the growth of an expressive vocabulary accompanied its very development. Monteverdi already early in the seventeenth century and Christoph Bernhard by the middle of the century, for example, recognized that certain affections require appropriate musical expression (Bukofzer 1948: 383). Yet, as noted above, expressive musical contents could still count as "discoveries" of what is inherent in music, rather than as musical idioms created for expressive purposes.[8]

Even in the twentieth century, discussions dealing with the subject mostly overlook the *formation* of expressive vocabulary. Most of those who discuss the question prefer to deal with the symbolic status of expressivity in music rather than with its historicity. The two, we believe, are interrelated. Bukofzer, for example, considering Bach's expressive vocabulary as representative of the baroque era, stressed its intellectual function, while denying its psychological immediacy. The symbolic association, argued Bukofzer, rested primarily on comparisons, and was obtained stepwise. The symbols—"allegories" as Bukofzer called them—did not convey meanings in themselves (Bukofzer 1939-40: 9). Rather, meaning was superimposed on sensual impressions, requiring knowledge of the relevant deciphering code. Bukofzer thus implied that the musical coinage of the baroque era did not consist of expressions, but of representations of an analytic kind instituted by conventions. Curiously enough, Tomlinson, sixty years later, supports, in a way, Bukofzer's position with regard to the "incomprehensive correspondences" of baroque musical expression, though he arrives at his conclusions from a Fouacualtian perspective (Tomlinson 1999: 34-61; especially 47).

Musical Coinage Conceived as Allegories

Bukofzer's position rested on two kinds of evidence—the musical treatises of the period which deal with expression and the music itself. While it is true that baroque theorists dealing with the *loci descriptions* emphasized "structural correspondences," they ignored questions of meaning. Still, what Bukofzer actually quotes from baroque theorists reveals that they also aimed at imagery, not only at intellectual riddles; why else would Calvisius, for example, see fit to tell us that "music employs elegant fictions which put the matter before the mind, the eyes and the ears" (Bukofzer 1939-40: 20)? Investigating the music, Bukofzer felt compelled to demonstrate that all baroque music consisted of "a sort of indirect iconology of sound" and that it "lacked all psychology in the modern sense."[9]

Bukofzer's argument may be presented in the following way:

1. Most of the "allegories" pertain to ideas that have nothing to do with the passions; they refer either to kinds of "bodily" motion (falling, ascending, etc.), or to abstract ideas, including that of emotions (as contained in concepts such as the "Ten Commandments," "earthly action" and "heavenly atonement," and the like).

2. The same interval or pattern may allegorize distinct notions with different emotional associations. This is only possible, explains Bukofzer,

"because the musical allegory is not expressive, it does not convey its meaning through the emotions" (Bukofzer 1939-40: 9).

3. The text is a mediator, assuring the deciphering of meaning.

4. The conceptions regarding emotions in the Baroque do not entertain their dynamic nature. Indeed, viewing the passions as a group of "typified and fairly static attitudes of mind" could only yield an aesthetic of strict figures that hardly coincides with the way we generally think of the emotional life.

Baroque music does indeed abound in "intellectual allegories," many of which refer to the passions in an indirect way. Such allegories were by definition products of deliberate acts on the part of composers, and they were consciously planned and consciously deciphered. Their connection with the rhetorical manuals makes sense both from a historical as well as from a logical point of view: they consisted of a "code" that had to be fully explicated. More recent historical studies, such as Ursula Kirkendale's "The Source of Bach's Musical Offering," reinforce this view (Kirkendale 1980; see also Street 1987). They reveal how sophisticated the use of rhetorical figures was, how they succeeded in giving rise to all kinds of riddles and a high play of wit as manifestations of the "humanistic concept of a high art, intelligible only to the elite." The *Correspondirende Societät der musikalischen Wissenschaften*, of which Bach was an honored member, had high intellectual aspirations, we are told, admitting newcomers only on the basis of their knowledge of matters of rhetoric and other sciences (Kirkendale 1980: 136). In other words, their competence—familiarity with the nature of the code—had to be guaranteed first. From a philosophical point of view, Bukofzer's allegories are analogous to Black's "comparison metaphors," calling upon the listener to detect the ground of the intended analogy so as to be able to reach the original meaning.[10] Comparison metaphors, however, do not exclude other types of co-extensive metaphors; philosophers assure us that the simplest of texts abounds in metaphors of all kinds. Baroque music too, we believe, consisted of more than "allegories." Moreover, Bukofzer's notion of "sameness of pattern" (point 2 above) is dubious. For example, chromatic treatment, which he views as a sort of "sameness," may assume very different forms. From an emotional point of view, there is a vast difference, one may argue, between a descending and an ascending or a dispersed and successive chromatic scale. Cooke, among other scholars, also insists that other parameters, such as rhythm, timbre and volume, may characterize a given intervalic pattern more specifically (Cooke 1959). In addition, (point 3) the text does not only temporarily mediate between the sign and the signified but rather substantiates it, as we shall learn from our British writers. On the whole, Bukofzer, like others, entertained a false view of the baroque conception of the passions. We have already seen how complicated both Descartes's and Mersenne's theories were, and they were not the only ones that dealt with the dynamic qualities of the passions. In fact, it is their kind of thinking that was elaborated in the eighteenth century, to the point of creating an affinity with modern psychological theories.

* * *

Musical Excursion No. 3: Man's Wailing

We chose the concluding Choral of the first part of the *St. Matthew Passion*—"*O Mensch, bewein dein Sünde groß*" (O Man, thy heavy sin bewail), to discuss some of the above points and the baroque conception of the affects in general. Our short analysis lacks, of course, a semiotic and philosophical underpinning which might have given it a more valid standing. Nevertheless we shall try to hint at some relevant concepts and a possible theoretical framework, which should substantiate an exegesis of this kind. In any case, all the components that we shall call attention to are audible, though some of them must have been more immediate to Bach's contemporary audience.

Any semiotic consideration of a movement like "*O Mensch*" should take into account that it functions as a "pillar" in the overall structure of this colossal work. For indeed, the dramatic, expressive, and symbolic components of the movement are supported by a huge architectural edifice, which they, in turn, substantiate. Architectural and other spatial metaphors are central, we believe, to the baroque style at large.

The text is a Lutheran choral text, and so is the musical infrastructure, which follows the division of the two-stanza text as well as its internal phrasing. The congregation (in Bach's time), once the opening ritornello was over, must have recognized what was heard, and though overwhelmed by the overall gigantic setting which encompassed it, must have mutely participated in the familiar melody. A feeling of awe, mixed with elevation, is, no doubt, what Bach sought to instill in his listeners, especially those who "faithfully" followed the twists and turns of the *"Passion"* up to this point (Bach's intentions as implied by his religious environment is explicated in Chafe 1991). This experience is, however, enriched by further symbolic content, which is integrated into the overall structure.

Bach seems to have been constrained by several musical factors, of which the choral melody and its underlying modality were not the least. Though the minor mode, at first sight, seems more appropriate for the text of the choral, Bach, who was compelled to use the choral predetermined major mode, took advantage of the expressive range it enables. Like many baroque composers, he used it as a starting point for shifting back and forth between "darkness" and "light," "depression" and "salvation" (ready-made connotations of such shifts then as now). Bach chose the tonality of E for his larger architectural structure and for the keeping of the expressive flow. At the same time, this was correlated with the choice of flutes as the main obbligato instruments (accompanied by the oboes d'amora); Bach was well aware of their natural expressive range. The flutes, prominent from the very first aria ("*buß und Reu*" [pentience and remorse]), also lead the ritornello in the second one ("*blute nur*" [only bleed]), carry the expressive burden in the concluding Choral both rhythmically as well as melodically (most remarkable are the expressive leaps, up [sixth] and down [seventh]), creating tension and connoting desperation (last beat of m. 2). The musical space is further magnified through the antiphonal texture of the orchestra, which "echoes" the *cori spezzati* predominant through the entire *passion.* A sublime distance between the flutes and the bass is thus insinuated. "Sublimity" is indeed what later aestheticians could find in the piece.[11]

The relentless nature of the ritornello, incessantly played throughout the choral, bestows upon it a "transcendental" necessity, so typical to Bach, the Platonist: Human utterances emanate from the ritornello only to dissolve back into the eternity which it represents. At the same time, the combination of ritornello and choral creates a semblance of a choral prelude, associated with the instrument (the organ) upon which God plays "in his administration and providence" (See Donne's *Sermons* as quoted by Hollander 1961: 264).

The melody of the choral, hovering above the other voices (reminiscent of the opening chorus, though now the choral is within the same "musical space," not above it), arrives to its first halt in m. 18, resting on the submediant (seventh), a "false" cadence. It thus represses the bustling turbulent emotion of the other voices. This is magnificently achieved in the next few measures, where below the silent soprano the lower voices, supported by the pedal point (anteceded in former cadences, marked by line in the music example), are moving into b minor, musically painting the "*bewein*" and almost literally weeping. This cry, so potent, is suppressed again. Cognizant as Bach was of the economy of "passions," he releases the stored energy only in the second half of the choral, in "*Krankenheit*" [illness] (mm. 61-67) and even more in "*für uns geopfert würde*" [for us he should be sacrificed] (mm. 75-85). He finishes the choral, and the whole first part, sotto voce—a level of dynamics that enables him to build the expressive impetus anew in the work's second part.

There is of course much more to this choral; the few points we raised exemplify, however, different semiotic levels and strategies—metaphors of various kinds, metonymies, entrenched connotations, occult associations, expectations and intertextual references. All of these function, of course, beyond contrived allegories. What is even more important, all is "orchestrated" to create a unified effect, meant to represent and to evoke a concerted disposition.

* * *

Kivy in his *Corded Shell* (1980), somewhat reminiscent of Bukofzer, speaks of the iconic nature of baroque music and argues that it involved an identification of certain contours with certain labels. According to Kivy, recognition, rather than sympathy or emotional impact, is the proper aesthetic response to this music. But Kivy makes no room for Bukofzer's classic-romantic inherent musical signs which, according to the latter, rest on a sort of natural connection between the sign and the signified. Whether inherent or superimposed, the attribution of emotion to music is cultural, and is not arrived at by whim. Once achieved, it may indeed function, as an entrenched symbol, rather than as a temporary parable.

Musical Example 3.4

Johann Sebastian Bach: "O Mensch, bewein"

Musical Example 3.4 continued

Alternative Views Concerning Musical Coinage

Still, there are emotional components, overlooked both by Bukofzer and Kivy, which exploit permanent cognitive resources. Bukofzer may have ignored them, because of the desire, at the time, to tighten musicological research and to avoid the kind of premises that accompanied musical research represented by scholars like Pirro (1907) and Schweitzer (1908). Bukofzer did find use, however, for the musical inventories created by these scholars, though they were not kindly received.[12] Schweitzer's preliminary attempt to sketch the elements of Bach's language does contain an important intuition: "The unique thing in Bach," he writes, "is the clearness and completeness of the language.... We can really speak of the roots and derivations of his language... These well-defined groups comprise, for example, the step motives for the expression of firmness.. the syncopated themes of lassitude; the theme that depicts tumult; the grace wavy lines that depict peaceful rest... the motives of rapturous, naive or passionate joy; the motives of distressful or noble grief." Though Schweitzer speaks of distinct groups of expression, what he actually describes are pairs of contrasting passions, expressed by contrasting musical means. Whether the means he describes are most representative or not of the emotions to which they were attached, does not change the fact that he tried to point out to an underlying mechanism of an expressive language. Moreover, Schweitzer recognized that "Bach's musical language does not consist in any special multiplicity of themes and motives, but in the manifold shading by means of which a few general formulae are made to express characteristic ideas and feelings."[13]

A similar approach, even more explicit, was later suggested by Cooke's conception of expressive music (Cooke 1959). Cooke considers the basic musical vocabulary in terms of major and minor chords and ascending and descending contours. These correspond to emotional dimensions, not to specific emotions: Whereas the "major" complies with "positive" emotions ("optimism," "victory" etc.) the "minor" complies with "negative" ones ("grief," "melancholy" and the like). The contours symbolize emotional dynamics: "ascending" contours convey rising tension, "descending" contours—their calming down. The form of the arch is, as expected, a combination of the two. More systematically than Schweitzer, Cooke suggests that idioms of expressive music imply a system, as it were, that enables their "location" within one and the same theoretical space. Such space comprises two basic dimensions of which the first relates to the *kind* of emotion (positive or negative) and the second to its underlying *dynamics* (growth or decline). He also entertains an explanation of the emotional "shades": time, timbre, and volume—the "vitalizing agents," as he calls them—which qualify the basic expression through and through.

Interestingly, a more recent study that deals with Mozart's rhythmic gestures implies a similar model. Duple and triple meters, the most common meters since the Baroque, became associated in the period, Allanbrook (1983) shows, with "exalted" and "terrestrial" passions respectively. The changes of tempi (themselves related to the binary "slow-fast") and internal divisions and stresses further qualified the basic duality. The classical composer, it is suggested, exploited these binary relations (and the continuum that they comprise) in order to dramatize the contrast between affects.[14]

Robert Hatten's recent study *Musical Meaning in Beethoven* (1994) posits an opposition at the center of his theoretical approach. Hatten views "markedness"—a term he borrows from Shapiro (1983)—as a concept standing for any differentiation which is distinctive enough for creating an opposition. However, in contrast to earlier writers, Hatten views this evaluation of difference as asymmetrical, in that one of the terms of the oppositions about which he is speaking, is always marked, whereas the other is unmarked. This, for example, is the case of the minor vs. the major mode, which, to Hatten's view, signifies a more defined range of meaning, compared with that which the major signifies. The same holds true for the standing of the dominant seventh in relation to the dominant fifth, or the diminished seventh to the dominant seventh. Moreover, Hatten stresses growth and flexibility as central to musical marked values: As new structural circumstances within a particular work result in new oppositions, the scope of stylistic markedness increases. At the same time, each musical entity may be a member of several different oppositional pairs, according to the context in which it is located. On the whole, Hatten views markedness as central to musical semiotics, for it enables us to overcome the problem of the non-referentiality of music by allowing us to probe the relative range or specificity of musical content in terms of dynamic structural configurations (Hatten 1994: 34-66).

Notwithstanding their differences, these various approaches to "music and meaning" share the intuition that musical expression, since the Baroque, consists of binary oppositions and the various spectra they imply. Empirical research into musical expressive language, and into the expressivity of the musical parameters in language rests on similar assumptions (Sundberg 1982). None of these studies, however, clarifies the theoretical gains achieved by such cognitive-aesthetic explanations. Interestingly, Webb, a central figure in our British group, based his entire theory on such assumptions, explaining thereby the correspondence between musical effects and the structure of affects, that is, the nature of the medium and the propensities of the mind.

The issue, however, is not purely theoretical; it is related to specific historical moments. At its best, aesthetic and cognitive constraints explain the underlying mechanisms leading to the emergence of particular expressive coinage. Cooke's argument, for example, that the different tonal tensions (embodied in the various major and minor configurations) "coincide," to use Webb's term, or "correlate," to use Hatten's term, with emotional tensions, is an observation which implies, however, the tracing of ways by which such correlations became entrenched.[15] That the elaboration of musical coinage should coincide with the preference for the aeonian and the eolian modes over others for the development of harmonic tonality deserves indeed careful examination.

The Evolvement of an Emotive Vocabulary

Any attempt to deal with the ways musical patterns became loaded with fixed meanings is problematic. The problems entailed are related to the condition of music as "the fine art of variations":

1. The realization in music of what philosophers understand by type/token relations is dubious on both the syntactic and the semantic levels. In language a token of a general type, say "table," is embodied in particular utterances denoting or connoting a token, that is, a specific table, whereas in music particular utterances, that is, tokens, do not directly refer to a pre-established type. In other words, in music motifs and their variants are not epistemologically distinguishable; moreover a variant can be affiliated with a new

motif, that is,a new type, since motifs undergo transformations in the course of the musical unfolding and may generate new configurations. Such bewilderment on the syntactic level seems to thwart clear and distinct associations on the semantic level.

2. The context in which a particular motif or component appears may change its meaning more radically than in language. Semantic types can be maintained only through literal repetitions of a work whose identity is guaranteed by the Western notational system.

3. Also, stylistic changes may affect the character of a basic musical idiom and thus change not only its meaning, but also its very identity. This condition is reflected in the treatment of musical styles as separate "languages" entailing different vocabularies and grammars.

4. The hermeneutic effort involved in interpreting a piece of vocal music is complicated for it entails text, music, and text cum music analyses.

5. The overall import of a piece should not be identified with the significance attached to its central motif(s) or to a simple combination of several of them. Combined, they are "troping" each other—qualifying and transfiguring initial meanings, as Hatten (1994) has argued.

To ascertain a theoretical basis for musical communication that would also allow for the uniqueness and the complexity of individual musical messages is complicated indeed. This stems, as we shall subsequently argue (relying on our British writers), from the fact that epistemologically speaking, music is constructed by *impressions*, rather than *ideas*.

* * *

Bearing the above in mind, we shall try nevertheless to exemplify the transformation of a certain musical pattern, one which received special and conscious attention at the time. It was called by Christoph Bernhard (1628-1692) "*Saltus Duriusculus*" ("poignant" leap), about which he says the following: "The leap of a minor sixth was not in use among composers of former times, except between re and fa, or between mi and fa. Today, however, the following are also allowed."

and he continues:

Musical Example 3.5a

Not as good

"Ascending and descending leaps of a diminished fourth, and descending ones of a diminished fifth are also permitted nowadays. The leap of a diminished seventh, which consists of a minor sixth and a major semitone, is also occasionally encountered, although almost exclusively in music for solo voice, and only in descending."

Bernhard apparently discusses a new melodic rule, rather than an expressive pattern, especially since it treats only one melodic step, neither rhythm nor harmony, and even that step is indeterminate. However, Bernhard, discussing the *Stylus luxurians communis* at this point, does intend to propose new expressive idioms which have become "permissible" in dramatic occasions of various kinds on behalf of their expressive power.

His example completes what is missing in his rhetoric:

Musical Example 3.5b

The text in the above example, we assume, imparts the betrayal of a lover. The "saltus," we further notice, occurs between a strong and a weak beat, and is resolved, in an embellished way, on the tonic. This is, in fact, the basic motif of Orfeo's *lamento* quoted in musical example 3.2 above. It exemplifies Bernhard's indeterminate position as to the size of the interval, for the diminished fourth is enlarged, when the utterance is repeated, to a minor sixth, as we have seen, thus exemplifying our point no. 1 above.

This was not the first time Monteverdi used this pattern. It already appeared in an earlier madrigal, "*O come gran martire*" where the leading tone is repeated several times, though without creating a dissonance (see musical example 3.5c; quoted by Fenlon 1985: 271). In both pieces, however, it is the opening notes, which are so handled; being a quasi-cadential figure, this pattern seems to impart "a sense of an ending" even before the piece has begun. In both cases, it serves the composer's specific needs, in the first case it imparts the notion of fatality and absolute grief, in the second, the poignancy and intensity of martyrdom.

Musical Example 3.5c

Claudio Monteverdi: "O come gran martire"

Pamina's *Ach ich fühls* from the *Magic Flute* (musical example 3.5d) is apparently related to Monteverdi's basic pattern only through their use of the g minor—a tragic mode, a *topos* unto itself. Actually it is based on and saturated by the same *saltus duriusculus*. No *saltus* announces the beginning (when a "drooping" figure fills the gap of the opening perfect fifth), but the first whole phrase moves between the focal points of D, G, F# a pattern which is immediately echoed by the tail-end of the phrase on the dominant [A, C# and D]—"*ewig hin der Liebe Glück!*"—in which it occurs twice, first on a diminished fifth and then on a diminished seventh (mm. 5-6). The dramatic context is similar to that of the former examples—morbidity is related to a false heart or to a cruel fate. This figure is no less prevalent in the rest of Pamina's aria, it embodied her cry: "*Sieh Tamino! Diese Tränen* ("mm. 17-18), echoing the piercing *saltus* of the diminished seventh of the flutes and oboes. As with Monteverdi and Bernhard, it is constantly extended to contain heightened expression. In the end it is stretched into a diminished twelfth, which is mitigated by a closing diminished fifth *saltus*, leading to the tonic (mm. 33-38).

Similar procedures seem to characterize the formation of other parameters: meter, rhythm, texture, timbre, harmony and their various combinations. The making of a musical vocabulary was dependent, on the whole, on textual and contextual conditions that gave rise to its differentiation in terms of emotive designations. Working within broad stylistic frames—*ecclesiastico*, *rappresentativo*, *concertato*, and the like—composers determined the qualities of religious vs. secular, lyric vs. dramatic, pastoral vs. military, pathos vs. bathos genres and many more. Specific musical contours, varied rhythmic configuration, certain harmonic progressions etc. crossed genres, giving rise to the intersections of ranges of meaning, a stage which the visual arts arrived at more than a hundred years before. By the eighteenth century, musical coinage covered a whole gamut of emotions and sub-emotions.[16] Once emotive communication gained "immediacy," the music itself could function as a label for expressive states without text, pretext or specified context. It is this state of affairs that is reflected in the so called *Affektenlehre* treatises, those of Andreas Werckmeister (1645-1706), Johann David Heinichen (1683-1729), and Johann Mattheson (1681-1764) and others, musicians and theorists who believed that specific expres-

Musical Example 3.5d

Mozart: "Ach ich Fühls"

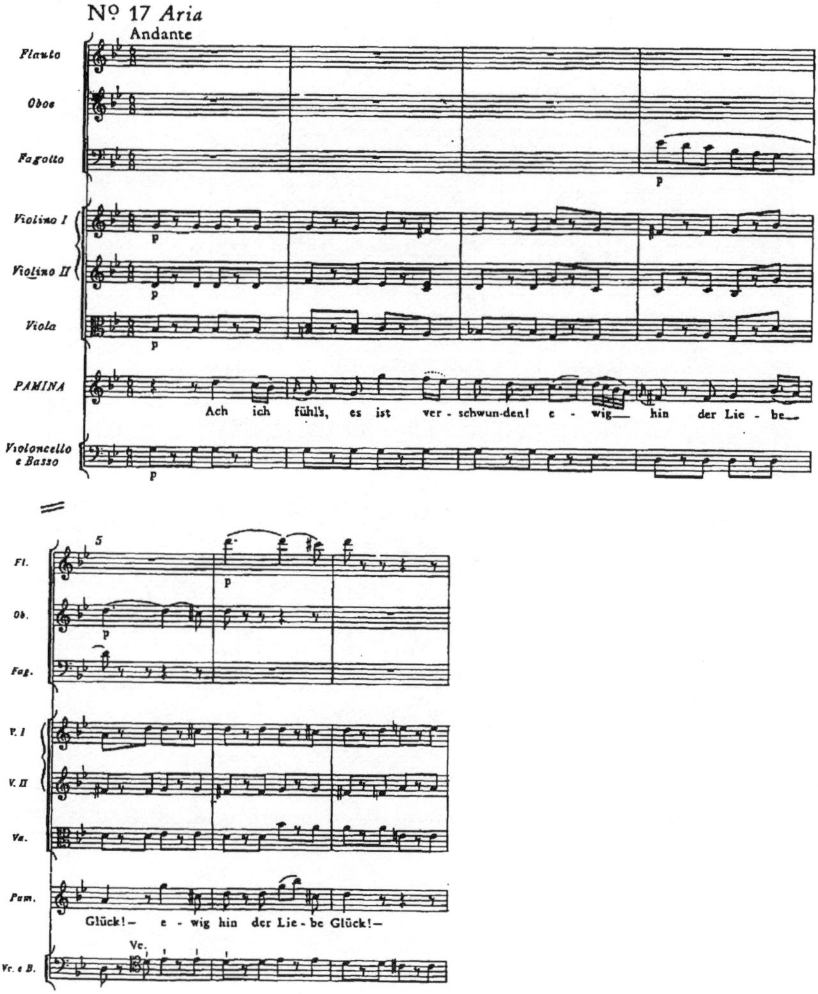

Musical Example 3.5d continued

Musical Example 3.5d continued

sions inhere in the musical forms themselves (Dammann 1967). Towards the end of the eighteenth century, however, with the change of intellectual climate, writers became aware of this fundamental fallacy, viewing musical semantics more and more in terms of learned schemata.

Generally speaking, the musical coinage created in the Baroque was basically of two kinds: (1) The "intellectual" kind—what Bukofzer called "allegories"—which consisted of "riddles," the solution to which required knowledge of the "code," rather than competence in a given style. Interestingly, this kind of musical coinage gradually dissipated towards the end of the era.[17] (2) The "emotive" kind, which was expected to affect the listener without intellectual mediation, that is, without the appendix of a deciphering code, for the "message" was built into the music itself. This explains why the "emotive" kind lasted as long as it did. Both kinds of symbols initially needed a text, yet it is only the second kind that came to "possess" the labels to the point where it could either do without them or even lend specific qualities to new texts.

Harmonic Tonality and the Representation of Passions

As already mentioned, the formulation of a musical coinage went hand in hand with the crystallization of harmonic tonality. Though the latter seems like an independent musical development, it was, in fact, part and parcel of the evolution of the newly forged symbolic dimension in music. In a way, harmonic tonality was another product encouraged by the experimental relationships between text and music ushered in by the opera. While each stage in the crystallization of the harmonic language reflects the symbolic potential of music at the time, the course of the entire development must be considered as determined by the will to endow music with communicative powers, independent of text, highlighting music's expressivity.

Viewed unto itself, the development of harmonic tonality is amazing. It started with ad-hoc harmonic relationships and ended with a thoroughly systematized scheme. Trying to portray affections, the early monodies were incoherent from a purely musical point of view. This incoherence resulted from a misunderstanding of the symbolic nature of art, a point we have already raised. In their attempt to be affective, the monodists saw fit to violate musical rules, so that the musically competent listener was often left with entities mostly cemented by extramusical factors.[18] Artistic intuitions, however, were stronger than dogma; the basso continuo, which "rescued" the new style by granting it a symbolic aspect, also proved efficient as a vehicle contributing to a new coherence. The artistic features on which the symbolic dimension in art hinges are often among the more organized; the relationship the basso continuo came to establish with the solo voice, did eventually determine the values of the new musical grammar. Distinct chords, connected to each other by diatonic, chromatic, or enharmonic voice leading—the "secret code" of the madrigalists—gained syntactic saliency. That the relationships among chordal configurations were important to composers is evident from the numerical figures that specify them, defining each configuration in relation to its adjacent. The new awareness of chordal configurations was not, however, embedded in a comprehensive system that treated chords as primary compositional entities. Composers made clear, nevertheless, that "points of arrival" should be governed by chromatic steps, while diatonicity should remain the norm for other passages. Chromatic and enharmonic modulations were preserved for dramatically loaded moments. The main dramatic tension was in fact created between the solo voice and the accompaniment, which could either conform to, or clash with, each other. (See musical example 3.2.)

The musical unfolding of such moments was accompanied by the increased tendency to exploit the contrast between the major and the minor modes. That the monodists should select the modes that gained theoretical standing only in the course of the sixteenth century is worth noting.[19] The tendency, since the fourteenth century, to move chromatically towards the cadence, granted superiority to the aeonian mode, which comprises the uppermost semitone diatonically. The monodists enhanced this tendency, emphasizing thereby points of arrival, as mentioned above. Taking into account that they also made use of the eolian mode, for contrasting effects, one may conclude that they basically opted for one mode, alternating its major third and sixth with their minor counterparts for expressive purposes, as Cooke (1959) or Zuckerkandl (1973) would argue. This choice had significant cognitive value, as we learn from present studies, making the most of tonal tensions, enabling contrasting expressions as well (e.g., Balzano 1982; Watkins & Dyson 1985).

The next stage of the development seems like a negation of the previous one, for the rhetorical basis on which intelligible settings of texts was built, was short lived. The Italian predilection for melodrama, for beautiful singing and patent forms soon gave rise to a musical setting that was concerned with the general spirit of the text rather than its details (Katz 1986: 107,174). However, the "general spirit" was realized by incorporating the new musical means. Graceful tunes and vocal ornamentations were now laid upon an infrastructure which made it possible for composers to match chordal configurations with stanzaic forms, subjecting the former to regular meters, coherent melodies and symmetrical structure. The different musical constellations were thus invested with values of poetical articulation. Divergences from a given tonal frame were short and did not stray afar; bold modulations were rare. "Rudimentary harmony," as Bukofzer (1948) called it, was the product of a stage of development that was interested in overall coherence, ignoring expressive subtleties. This resulted in more clearly defined diatonic harmonic relations, while chromatic and enharmonic connections were excluded, though not totally rejected. In contrast to the more lyrical parts, recitatives preserved some of the "ad hoc" harmonic procedures that characterized the early monodies. In the old homophonic settings—the frottola, canzona and the like—chordal formations exist, but they were not engendered by root progression. It was only when chords received major attention that their functionality became gradually systematized.[20] The explicit enumeration of their intervalic structure guaranteed the correct linkages among them, in terms of voice leading, enabling extensions and contractions in accordance with the poetical structure (musical example 3.6).

* * *

Musical Excursion No 4: Effective Incanctation[21]

Simple, chordal 3/4 accompaniment on E minor, prefigures, through repeating, measured, pauses, the horrible caverns of Medea's vindictive energies from Francesco Cavalli's *Giasone* (1649). Its monotonic character serves as an appropriate foil for her simple, syllabic, song. Using the range of an octave (in the first verses) and the tenth (in the last verses) for ascending and descending arpeggi, the tragic heroine and the accompaniment oscillate between the tonic and its relative major (C), showing another aspect of her militant character. However, the dominant (on B), on her penultimate measure, and the related leading tone (on D#) creates strong cadence on the tonic. Notwithstanding the simple structure, each of the six-measure verse stands for a distinct embodiment of its underlying speech act.

* * *

Altogether, it is due to the internal syntactic coherence of chord progression, which the previous stage lacked, that melody could not only return to the musical setting, but also interact intimately with the musical texture as a whole. Two major rules underlie the melodic perception of Western listeners, according to cognitive studies—the intervalic order of the diatonic scale and the relative frequency of the notes that are the nearest to the tonic in the cycle of fifths (Dowling 1982). The new harmonic settings, one may argue, created these basic conditions also in the vertical dimension. If previous works had based harmonic relations and contrapuntal texture on the dictates of a certain *cantus firmus*, now they shared the same underlying grammar. What transpired in the field of harmony parallels, in a way, the transition from "melody types," in the chant repertoire, to modal organization, which freed melodic thought from predetermined patterns.[22] In the case of harmony, separate entities became unified and hierarchized, enhancing directionality in a system of expectations that facilitated perception and enabled, paradoxically, greater variety and change.[23]

The musical potential contained in this new state of affairs was of high consequence. For the first time in music history, the melodic and contrapuntal elements could truly experience a "complete union in the joy of music-making" (Bloch 1985: 4). The new possibilities, however, were not immediately realized. During the first half of the seventeenth century, when rudimentary harmony became already the rule for vocal compositions, instrumental works—such as the ricercar, toccata, and canzona—still used modal tonalities and contrapuntal procedures of the old type. It is not a mere coincidence that those who tried to integrate the *stile osservato* (the strict style) into the new system were composers who were also active in the dramatic field (e.g., Alessandro Stradella, Giovanni

Musical Example 3.6

Francesco Cavalli: "Virago incantation"

From: Arnold Schering, Geschichte der Musik in Beispielen: Dreihundertfünfzig Tonsätze aus Neuen Jahrhunderten, Wiesbaden: Breitkopf und Härtel, p. 252.

Legrenzi). Their instrumental works became "harmonized," their texture more transparent, giving predominance to the "singleness of melody," as Rousseau eventually termed it. Most important, the old imitations were now governed by the harmonic scheme, submitting academic procedures to a more immediate developmental process.

Without poetical dictates, composers were now able to exhaust harmonic potentialities. Oriented by imitative techniques, they translated them into motivic elaborations, which govern melodic as well as contrapuntal procedures. In the domain of a given tonality, they discovered melodic sequences, based on passages of sixth chords, to be most compelling. Elaboration of thematic material was furthered through modulations. Composers seem to have opted for "variegated sameness"—texture governed by permutations of the same basic thematic material (modified melodically as well as rhythmically), highlighted by the different harmonic and tonal frames. The old counterpoint became newly available, enhancing the new motivic coherence, while the directionality implied by harmony was realized not only in "continuous homophony"—the most common kind of texture in the late Baroque (Bukofzer 1948)—but governed as well contrapuntal textures, including that of the fugue.

By the time of Corelli, the harmonic system achieved coherence on two interrelated levels. On the micro-level, which refers to the underlying tonality at given points, each musical entity became defined in terms of the tension it created, measured by the distance of its root (in the cycle of fifths) from the local tonic. On the macro-level, whole musical "areas" gained their coherence according to similar criteria, referring to the overall tonality of the piece. Each musical entity was thus doubly defined via basic rules that applied to both levels. The definitions themselves, however, were contextual; the same chordal constellation could belong to several keys, enabling modulations with no fixed rules for their occurrences. It supplied a new resource of expectations, frustrations, and surprises that could operate on a large scale (the above Bach example [3.4] illustrates all these points).

In retrospect, one may say that the new harmonic system entailed properties of high cognitive advantage: It entailed (1) two levels of organization working through a shared scheme which (2) unified the vertical and the horizontal dimensions by (3) rules of coherence and "directionality." These properties amount to a "parsimonious description" of the sensational array which psychologists consider highly important for the improvement of musical perception.[24] Moreover, implied sets of expectations, one may argue, increase the possibilities for extended works because of their efficaciousness regarding memory. From a compositional point of view, it enabled the creation of a variety of textures and techniques distinguished by melodic predominance, motivic elaboration, rhetorical structure, and clear contrapuntal strategies. Sharing the same system, musical textures could be arranged in juxtapositions, allowing for gradual changes, contrasts and similarities, which could be exploited for dramatic and expressive purposes. In the final analysis, all of the advantages of the new system stemmed from its potential to constitute a powerful referential scheme. Previously, each musical piece had to create its own internal order, employing specific rhythmic patterns, strategies of melodic imitation, contrapuntal devices of sorts, and so on. With the new system, there was no longer a need to expose the whole network of relationships in every composition, for it gradually resided in the mind of the listener constituting an unconscious frame of reference. This, as mentioned earlier, also improved musical retention, making the enlargement of musical works possible without damage to coherence. The "continuous expansion" of the High Baroque, as Bukofzer indicated, was "a formal principle that lent itself to infinite variation as to formal patterns" (Bukofzer 1948: 359). This would have been inconceivable without the cohesiveness of the referential system employed. In fact, since the end of the seventeenth century, the system has functioned like an "overlearned alphabet"—to use a concept from perception theory, facilitating the apprehension of Western tonal music.

Before the harmonic system was fully developed, composers used *motto* beginnings in their arias, intensifying the symbolic relationship between the solo and the accompaniment. The latter, while ushering in the solo, revealed the protagonist's state of mind, so to speak, before he himself was ready to express it. At the same time, temporal structures became less restricted, allowing for prosaic rhythms within metric structures. At that point, the dependence of music on text, which characterized the *bel canto* style, gave way to the predominance of music. As the overall harmonic structure became more defined, the music gained in autonomous standing. Opting for the "variegated sameness" of the new style, composers and librettists intuitively shortened the text, so that by the time of Apostolo Zeno it came close to being a "label" for an emotional state.[25] An altogether new relationship between text and music ensued, a relationship appropriate for the expression of affects, whether single or mixed.[26] Webb, Smith, and some modern philosophers, especially Goodman, will be called upon later to help explain the symbolic significance of such a relationship. In the meantime, it is important to note that the dramatic works of French and English composers did not exhaust this symbolic potential as did their Italian counterparts. History, however, opted for the Italian way, Joseph Addison and St. Evermond notwithstanding.[27]

To paraphrase Webb, vocal music, "borrowed" an elaborate grammar from instrumental music and "lent" it symbolic meaning. Monothematic elaborations of instrumental music were thus granted the metaphorical status of an elaborated psychological state of the kind explicated by Descartes. (This should not be confused with the specific symbolic meaning related to molecular musical coinage, discussed earlier.) Indeed, it is through exchanges between vocal and instrumental music that instrumental music continued to entertain the fictive status that was bestowed upon music in the early Baroque.[28] Solo and tutti became the "virtual agents," as Cone (1974) put it, of a metaphorical discourse in the concerto. The concerto, in turn, lent its fictive state to the articulation of other instrumental forms, including the fugue. It is important to note, however, that while new forms of expression were steadily developing, older forms continued to be practiced. In the final analysis, it is the accumulation of the possibilities to portray emotions that granted late-baroque works their expressive power: Whereas the *recitativo secco* enabled musical transmission of "dry" or "narrated" information, the *recitativo accompagnato*, with its bold melodic curves and harmonic transitions, could effectively highlight soliloquies. Choral parts could be enlisted for "contemplation" or for the realistic evocation of crowds. It was the spirit of the *da capo* aria, however, which dominated the "scene."[29] No wonder then that it was the aria and its derivatives that mostly intrigued later thinkers when they came to ponder over the relationship between music and poetry.

Musical Narrative: the Classical Unfolding

The developments discussed above amount to a major paradigmatic change in Western music that held sway over three hundred years. Most of the developments during that period were, in one way or another, the working out of the implications of this paradigm. Considering the classical style as such a "working out," rather than viewing it in terms of a "break," amounts to an examination of the ways in which the new fictive status of music, its expressive vocabulary and its basic grammatical structure, were exploited.

The stylistic trends in eighteenth-century music are often described as governed by a drive for clarity and intelligibility characteristic of linguistic discourse. These trends, it is argued, culminated in the aspiration of the high classical Viennese style towards the dramatic. By contrast, some see in it precisely the demise of "language oriented music," at least in its last phase. According to Neubauer (1986), for example, the tradition of imitation and expression was actually broken during this phase, while an orientation towards purely formal organization ensued. The new music, he argues, heralded a new aesthetics with a Pythagorean bend. Whether linguistic or mathematical, it is agreed that instrumental music became "emancipated" by the second half of the eighteenth century; it no longer required the assistance of text to get the attention and honor it deserved. It is to this emancipation, Dahlhaus (1989) tells us, that we should attribute the eventual idealization of instrumental music expressed by people like Wackenroder and Schopenhauer.

What made this emancipation possible? Was it achieved through transformations of linguistic procedures or through the severance from them? Lines of arguments are often confused; even the term "absolute music," argues Treitler, served contenders of both sides "like a political banner that is seized by persons of quite divergent persuasions" (Treitler 1988: 411). Treitler himself tried to show that it was through a realization of mental processes, related to narrative discourse, that the High Classical music exerts its effects. Yet he did not fully acknowledge that music was able to appear emancipated from poetry only as a result of its previous inculcation of poetical procedures. In what follows, we shall argue that it is this inculcation, which is metaphorical by definition, that paradoxically enabled the emancipation of music. Moreover, we shall contend that it is the poetical metaphor, which highlighted the "condition of music."

The following discussion of the classical style is thus guided by the stages of the formation of a metaphor. We have discerned three major stages in this development: (1) the making of a "syntax," (2) the creation of a "narrative structure," and (3) their investment with "meaning." The first two stages concern the structural aspects of music, while the third calls attention to processes whereby the new structures merged and interact semiotically with old iconic meanings, furthering the cohesion of the structural elements. Historically speaking, the first two stages occurred, more or less, simultaneously, between 1720-1760. Twenty more years were needed to combine the two significantly. The last stage, which took place mainly in the eighties, crowned the entire development, endowing music with the kind of coherence that eventually became identified with the classical style.

The Scaffolding of Classical Syntax and Pragmatics

Within the modest limits of the Neapolitan operatic overture, a new kind of musical articulation emerged: the *scaffolding*, as Helmut Hell (1971) calls it. It was abstracted from a certain fanfaric formula used to herald the forthcoming opera. The starting point of the new musical articulation was thus an extended arpeggio, metrically articulated. It relied on the internal tension between the two components ("cells": a' and b' in musical example 3.7a) of the basic unit of the fanfare—each consisting of two beats—and on the tension between the units (of four beats) comprising a "block" of units which, altogether, constituted the unfolding of the arpeggio—the basic fanfare (example 3.7b). Hell, relying on Georgiades's studies (1950; 1951), shows how these basic elements functioned as a "scaffolding" for the entire opening movement of the overture (see example 3.7c; each block marked by a capital letter). From his description it appears that the two-beat cells, the four-beat unit, the building blocks, and their extension into larger sections were held together by: (1) a slow harmonic rhythm, determined by the metric-melodic interdependence between the basic unit and its encompassing building block; (2) motivic components defined by the cells (fitting the length of two beats or multiples thereof); (3) sequential behavior determined by the first two units of the block to which they belong; and (4) "gaps," implied by pauses occupying the second cell of a unit, striving towards antiphonal completion. (In musical example 3.7c, unit b in block A is in fact a completion of such a gap [created by the strings] by the oboes, horns and trumpets.) (5) Multiplications of basic unites, which preserve, on a new lever, the same structure while hierarchically containing the lower level (block B in musical example 3.7c). This adds to the acceleration effect. On the whole, the striving towards climatic completion is part of a larger "drive" towards the equilibrium that characterizes the system (see the ultimate block in musical example 3.7c).[30]

Musical Example 3.7a

Scaffolding - basic bricks and fanfaric layer

Additional constraints refer to the function of the blocks following the first block and to the nature of tonal extensions. Accordingly, the "climbing" nature of the first block, its ascendance towards a melodic climax, affects the other blocks or phrases. The latter can either further (block B in example 3.7c), hold and large (block C) or retard (block D, despite the modulation, because of the breaking to single "bricks") the tendency of the first block, though they are all designed so as to further the overall climatic structure. Climaxes can be achieved through textural, orchestral, rhythmic, and dynamic factors, in addition to the melodic one. Special blocks are needed to convey a closure or a new beginning, yet they, too, should forward the thrust of the whole. The extension of tonal areas, on the other hand, is basically determined by the fact that the first block consists of an unfolding of a single chord, along it contrapuntal texture. This furthers the unity of each section (cf. Pergolesi's example with the opening of Mozart's overture of *The Marriage of Figaro*; of the exposition of his Piano Concerto in C, K. 503; for clear examples of similar dynamics).

Taken together, these constraints affect not only the dynamic character of the various sections, contributing to the effect of the whole, but also to its cognitive efficiency, through their limitation of musical procedures associated with the high Baroque. According to modern psychologists, perceptual efficiency in linguistic systems is bought at the expense of variety and flexibility (Watkins & Dyson 1985). Specificity and conclusiveness (inflexibility) are consequences of the rules governing the "well-formedness" (to use Chomsky's phrase) of grammatical systems. It should also be noted, that "linguistic freedom," according to Chomsky, is organized by what he calls "transformational rules" (Chomsky 1957). Naturally, the "transformational rules" of the "scaffolding system" could only be developed once the rules related to "well-formedness" had become entrenched; it did not take long until this transpired.

Both rules of well-formedness and transformational rules may be found, in different degrees, in other musical styles as well. The novelty of the "scaffolding system," however, was that the unfolding of an entire structure was implied by a kernel—the basic unit—which determined the principle as well as the subsidiary clauses (blocks). It is this generative phenomenon that granted the system its linguistic semblance. Yet, while the syntactical structure of natural languages and the speech acts it give rise to rest on a composite scheme of relationships, involving various correspondences between words and real or possible states of affairs, in music, by contrast, all syntactic rules, are, first and foremost, internal. This does not mean that musical utterances are denied illocutionary force; instead of establishing reference through analytical analogies they rest on synthetic strategies.[31] Historically speaking, the desire, in the Italian overture, to simulate a "drive towards a destination" gave rise to the "scaffolding system" and not the other way around. The function of the Italian overture, as mentioned above, was to usher in the opera, to create a

Musical Example 3.7b

C. Pallavicino: "La Gerusalemme Liberta" (opening)

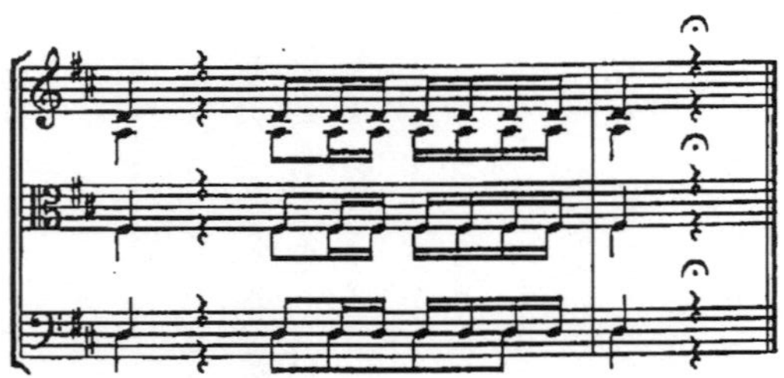

Musical Example 3.7c

G. B. Pergolesi: *St. Guglielmo Duca d' Aquitania* (overture)

zero degree layer

A a' b' Quasi antiphonal layer

Simple + double scaffolding

B

Double - antiphonal

C

D

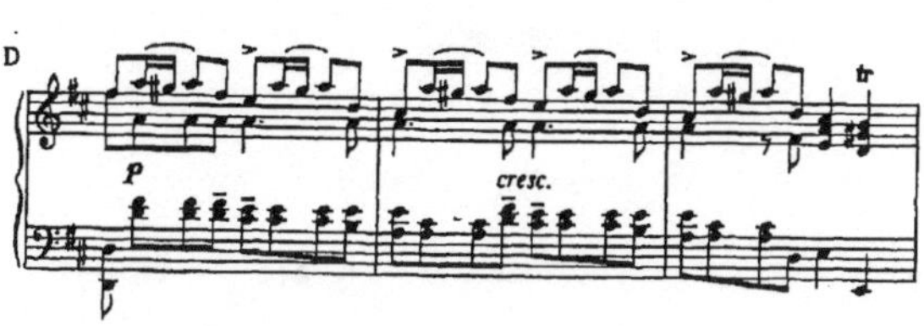

"Tutti" treading layer, single bricks

B

"X" area, from A major back to D major

Musical Example 3.7c continued

C

C'

D'

Mosaic recapitulation

C

D

Musical Example 3.7c continued

dramatic sense before the real drama actually began (Sondheimer 1922). Indeed, what was implied by the fast-slow-fast structure, established by Alessandro Scarlatti, developed into a well-ordered syntax under his Neapolitan followers, a syntax which guaranteed the intended forward thrust, furthered by its internal articulations.

In most natural languages, the construction of units larger than a sentence, we are told by linguists, is left to pragmatics, to ad hoc contextual considerations, implying certain notions of order and continuity. Pragmatic considerations also played a part in the "scaffolding system," related as they were to decisions concerning length of sections, the frequency of their occurrences, their density, and so on. All these, however, were more conspicuously governed by the same basic generative rules, than are similar considerations in spoken or written language. In other words, the "scaffolding system" behaved like a grammar governing not only the phrase, but the discursive structure as well. While the syntax of language is based on a preliminary level of semantics, the syntax that music evolved was able to create tensions, relaxations and the like, overriding the semantic level. This gave rise to new illocutionary values imparting the sense of a unified message, akin to the implicit mood created by poetical texts. The "mission," which music theorists since the sixteenth century were concerned with, that is, the systematization of the emotive part of rhetoric, was thus accomplished, though in an unforeseeable novel way. Contemporaries, who were aware of this development, tried to account for it through simple analogies with language, partially ignoring its metaphorical nature.[32]

The dissemination of the "scaffolding system" was rapid. It was incorporated early into the *parlando* style of the *opera buffa*, lending dramatic situations the support of a "rhetorical grammar." Outside the realm of opera, it was adopted by symphonists, among whom Stamitz is, perhaps, the most outstanding example. In his symphonies, the system underwent a further development through the introduction of "periodicity," which later became identified with the classical style.[33] Symmetrical phrasing was known, of course, through the strophic song and dance, long before Stamitz or any other so-called pre-classical symphonist made use of it. The emphasis on simple structures, in *opera buffa*, only increased the impulse toward symmetry. Still, there is a significant difference between, say, a Neapolitan song of the beginning of the century and a well-balanced period of the ripe classical style. In contrast to the former, the latter is hierarchically organized; it is this organization that renders the composition so powerful. The full-fledged hierarchical system of the period was, in fact, the product of a long process, we learn from Wolf (1981). The transformation of musical relationships from the lower to the upper levels was anything but automatic; the elaboration of each level necessitated a re-organization of the others, if one's objective was truly to achieve well-balanced musical phrases. Aware of the difficulties, Wolf, however, overlooked the origin of this hierarchical structure, that is,the "scaffolding system."[34] It was the latter that provided the basic musical materials "to begin with," as well as the grammar for their employment (See musical example 3.8 [from Mozart's C major String Quartet, K. 465, first movement] for how the two are interrelated).

* * *

Musical Excursion No. 5: The Well-Made Period

Basic conditions for a well-formed period would be:

1. A scaffolding infrastructure.
2. Basic division into two symmetrical complementary parts (A vs. B in musical example 3.8), which may be further divided symmetrically, without endangering the hierarchic closure. A variant repetition of this construct (A' vs. B' in musical example 3.8) may "endanger" this well-formedness (enabling its "reopening" in terms of the musical narrative).
3. On the higher level, the two parts—functioning as "antecedent" and "consequent"—are distinguished through a half cadence (m. 26), and an ascending vs. descending melodic contour, reinforced by a denser harmonic rhythm in the consequent.
4. On the lower level, the antecedent is made of two adjacent double or simple scaffolding units (see the antiphonal completion of the scaffolding basic "gap" in m. 31), sequentially related, and clearly separated from each other. The consequent, on the other hand, breaks the "blocking" nature of the antecedent with a more continuous melodic line (in our Example this break is concretized also on the textural level in the solo descending line of the first violin (m. 30).)

While opening a first (sonata allegro) movement with a well-formed period is quite rare in classical music (in Stamitz and the Viennese alike), it is explained in our example against the background of its famous twenty dissonant opening measures. This "periodic" denouement is short, in this Haydnesque Quartet, however, for already in measure 36 its closure is breached: basic scaffolding units are emancipated from their previous "fetters," ready for new adventures.

* * *

The statement that the classical period developed from the scaffolding needs no further elaboration, for it is evinced in each musical piece in this style. Less obvious, however, is the fact that *cantabile* periods were integrated into a "block-setting" texture. While they enriched instrumental music, they themselves were enriched by the new rhetorical possibilities. The well-balanced period, moreover, opened up new possibilities for large-scale balancing procedures, with implications for whole movements. Indeed, the setting of a movement was now enriched by varying types of closures, which allowed for growth.

The tension between the open and the closed, between the scaffolding blocks and the variants of the periodic structure, became a new source of drama. This kind of tension played a central role in the *Sturm und Drang* phase, granting sonata-form structure new dramatic powers. But already in Stamitz's time, the period, which allowed the instruments to "sing," assumed the status of "virtual agents" that was obliterated by the fanfaric gestures of the overture.[35] In short, if the "scaffolding system" made it possible for the sinfonia to preempt the drama, the drama itself could now be incorporated into instrumental movements.

The fictive dimension of instrumental music, however, was not achieved through its association with opera alone, since the periods functioned as variegated illocutionary statements, in relation to the regular scaffolding blocks. Rhetorical gestures of all kinds were now available to the composer; gestures that could be enlisted to fit desired utterances. The possibility of highlighting and suppressing musical events in relation to the whole gave rise to the narrative as a new metaphorical resource. The most essential property of the narrative, Booth argues, is the selective presence of the author's controlling hand. The author's voice, we are told, is revealed in every utterance, even when his "point of view" is momentarily hidden (Booth 1961). As we have seen, the composer's voice became salient as soon as the symbolic relationship between melody and accompaniment became clear. The musical symbolic representation of consciousness was itself a divergence from other dramatic displays; the "composer's persona"—to call upon Cone once again—was thereby logically substantiated.[36] It is the rhetorical constructions, facilitated by the new "grammar," however, that re-enforced its psychological presence. This presence, as is well known, became a main focus for the romantic composers, who were more than aware of the projective nature of their compositions.[37]

The Narrative Structure of Sonata Form

Important as the period was for evolving musical narrative, its impact might have been lost had it not been for sonata form. The

Musical Example 3.8

W. A. Mozart: *String Quartet in C*, K. 465, first movement

crystallization of sonata form was not as gradual and continuous as one might expect. Its appearance, in fact, took place in different places almost at one and the same time, in northern Germany and in southern and northern Italy. The transition from the binary structure of the suite movement, or the ternary structure of the *aria-da capo*, to the structure of sonata form cannot be explained without taking into account the conscious search for an alternative that would neither be merely architectonic nor simply decorative.

In the development of sonata form, the emphasis of each composer, of each school, was somewhat different,[38] yet the integration that eventually took place included most of their innovations. The most central innovations consisted of: (1) the division into the three sections which eventually became known as "exposition," "development" and "recapitulation"; (2) the subdivision of the exposition into two main tonal areas; (3) the enlargement of the tonal areas to permit their functioning as frames of reference; (4) the treatment of the modulation from one tonal area to the next as a central event (the "bridge"); (5) the harmonic and contrapuntal elaboration, in the development section, of the thematic material of the exposition (taking freedom from the "periodical" dictates but not from the "scaffolding" ones); (6) the resolution of the tension built up in the development and the harmonic preparation for the return to the initial tonal area; and (7) the return of the thematic unfolding of the exposition in the recapitulation preventing its tonal conflict.

Awareness of the metaphorical import of sonata form led many writers to linger on its affinity with drama and the dramatic;[39] only lately was attention paid to the narrative. Thus most writers emphasize the contrast, or even the conflict between the tonic area and that of the dominant. "Modulation in the eighteenth century," writes Rosen, "must be conceived as dissonance raised to a higher plane, that of the total structure," while previously it functioned as "a drifting movement...a cleidescope shift."[40] Concerning the development section, Kerman writes that "the impression is always of some sort of close struggle, hazardous exploration, terror, uncertainty, dramatic clarification, or trial" (Kerman 1952: 78). It should be emphasized that these and other metaphorical attributions, arise from the juxtaposition of the relatively relaxed pace of the exposition to the intense working out of the development.[41] While all agree that the recapitulation is no mere repetition, some suggest that it implies a kind of "remembrance of things past," that bestows a new status upon musical events.

These kinds of reading of the roles of the main segments of sonata form enforce the narrative as metaphor. Treitler, for one, seems to have understood these conditions and their implications. "Our word 'exposition,'" he writes, "carries the sense of direct presentation, but we should be explicit about recognizing the implication of the opposite, indirect presentation, in the 'recapitulation'—not only 'telling again' but also 'telling about.'" Discussing the allegro movement of Mozart's g-minor Symphony he says "the last presentation is like a reflection or conversation about the theme in the upper strings, above a lingering inflection of the tonic in the bass. It has the quality of a meditation over a remembered past, after the turmoil and before the silence."[42]

While many agree that the sonata gained power from the dramatic domain, it was never clearly specified what was borrowed and, more important, what were the cognitive implications of such borrowing. At any rate, once composers no longer thought of the different sections in terms of an ordered sequence, but rather as *functional* entities whose significance could be understood only in retrospect, the "plot" emerged. Viewing the plot of a story as a chain of interrelated functions is perhaps one of the most accepted notions concerning narrative; it enables the construction of a more conclusive theory of narrative, possessing greater explanatory power.[43] Sonata form, however, did not only share the functional aspects of the plot, it also shared the dynamics of its content. If indeed the essence of sonata form is *conflict, struggle,* and *reconciliation*, it does resemble what narratologists refer to as the "deep structure" of narrative.[44] While it may not have the epistemological standing that deep structure enjoys in linguistics, it does characterize a vast portion of Western narrative.[45] At any rate, it supplied a general frame of reference for composers and listeners alike.[46]

Entrenched as this frame of reference became, it involved, nonetheless, metaphoric interaction. When a "foreign" predicate is applied to a subject in language, the interaction between the two (frame and focus), takes place in the mind. Since music does not operate through labels, the interaction must take place within the "text" itself, that is, inside the music. That does not mean that the listener's mind is not engaged, only that the listener, in this case, is asked to supply the labels, once he processed the interaction. Moreover, whereas in a real narrative, the structure is independent of the "chronological" unfolding of the story (hence the division, in narrative poetics, between "fabula" and "sujet"), in a medium void of lexicality, order and structure cannot be separated. Sophisticated as the sonata form might be, it can never open with the "recapitulation," nor end with the "development." Narrative, or, for that matter, myth do *not* behave like musical scores, as Levi-Strauss would have it. On the contrary, that which is "contrapuntal," as it were, in narrative, that is, the extension of the plot, is arranged "melodically," that is, sequentially in music, and vice versa.[47] Generally speaking, metaphorical interactions among the arts do not imply simple similarities; "reversal of relationships" is, more often than not, the underlying rule.

On the whole, sonata form, like perspective in painting, is a convention that grew in response to new artistic needs, which entailed cognitive implications. Sonata form, as it were, educated the listener to link musical events in new ways; the listener was asked to compare earlier and later occurrences in ways similar to the novel. Eventually, such procedures became part and parcel of musical constructions as different from one another as a late Beethoven work is from a Wagner opera. In short, "conflict," "struggle," "reconciliation" and the like could, henceforth, take place in compositions other than those constructed by sonata forms.[48]

As a metaphorical construction, sonata form made salient, through independently recognizable patterns, what is implicit in the literary form. Writers were now free to "borrow" from music that which was lent to music by literature not so long before. Whether or not such a borrowing fostered the musicalization of certain novels, it undoubtedly contributed to the conceptualization of their structures.[49] Interestingly, it may be said that what was abstracted in music in the eighteenth century, was "discovered" in poetic theory only much later.

Narrative Invested with Meaning

Impressive as the early constructions of the classical style may appear, it was the Viennese who widened its scope and dramatic significance.[50] Reared on the *stile osservato* and on free contrapuntal techniques (Kirkendale 1979), their efforts were directed towards an integration of the prevalent musical language with baroque procedures. Paradoxically, it was this integration that made the full realization of the basic classical tendencies possible. It also gave rise to the "new way" developed by Haydn and Mozart in the eighties, which is inconceivable without the "revolution" that occurred a decade earlier.

* * *

Musical Excursion No. 6: Periodic Scaffolding of Dramatic Tension

Haydn seems to have been the first who exploited the tension between the scaffolding units and the periodic structure, realizing that it could enhance the overall dramatic impact of sonata form. His *Sturm und Drang* works are full of interesting experiments in this direction: he may open a movement with an imbalanced period (musical example 3.9a, mm.1-4), in which the basic units of the scaffolding appear as though they were trying to "free" themselves from the restricted hierarchy of the period. This becomes more prominent in the second phrase, which though only a variation on the first one, exposes the asymmetry and imbalance among its units. This "antagonism" soon results in a break of the outward symmetry leading to the modulation (mm. 14-5). The arrival in the mediante appears more like the outcome of a "crisis" than as a conventionalized procedure. The contrasting tonal area, compared to that of the tonic, is clear of such predicaments; period and scaffolding units live in harmony, without damaging each other's independence. The recapitulation shows that this approach is fully acknowledged by the composer; for he chooses to forgo the initial tension—the opening theme is changed so as to avoid the previous inequilibrium (musical example 3.9b). On the whole, one feels that it is through a particular handling of the music that the drama unfolds; by rendering it more concrete, it becomes more convincing. Different phrases are also more intimately connected, thereby enriching the overall rhetorical structure.

* * *

The intimate relationship between the scaffolding unit and the period constituted the first step towards the recognition of the importance of motivic unification to the cohesion of the drama. Already in the seventies motivic contents got more attention than before and became more expressive and more unique. Their expressivity derived from the fact that they were modeled after the baroque musical coinage discussed earlier. Emotional contents were needed to fill simulated "vacant" rhetorical frames created in the earlier phases of the new style. At the same time, after years of acknowledged rejection, contrapuntal techniques were let in, through the back door, into the last movements of extended works (e.g., in Haydn's string quartets op. 20, no. 2 ,5, 6). In these movements composers amused themselves with fugues and fugue-like techniques. In retrospect, such movements appear as contrapuntal exercises, integrating baroque procedures with classical structures and ways of thought.[51] The role of the antiphonal texture, embedded in the scaffolding system, was crucial in this respect.

This integration, which gradually invaded first movements as well, contributed to the metaphor its last clinching touches—form and content were significantly bound. Once jointly entrenched, they could function as a "natural" language for composer and listener alike. Accompanying parts were no longer expected to provide chordal bases alone. Not that the distinction between melody and accompaniment was obliterated; rather, the one was simply transformed into the other, as Rosen puts it (Rosen 1976: 17). As for contrapuntal techniques, composers emphasized the *discursive* aspects of the accompanying parts; they were, mainly, "responding," "reacting," and adding to the principal musical utterances. This broadened role of the accompaniment enhanced the fictive status of the entire metaphoric structure, in which "virtual agents," contributed to an ongoing "conversation." Motives and themes were more and more chosen for their potential as material to be elaborated and developed in accordance with an encompassing dramatic scheme. Special efforts were made to avoid standard and readily available musical materials, for the material had to be of a sort that "can be made to release its charged force so that the music no longer unfolds, as in the Baroque, but is literally impelled from within." According to Rosen, this was Haydn's greatest contribution to the history of music (Rosen 1976: 120). Following Hanslick, one could claim that Haydn's "new way" of the eighties amounted to a conception of music as "forms activated by *motives*." At any rate, the motive constituted not only a source of energy, but it supplied a recognizable physiognomy as well.[52]

Musical Example 3.9a

Joseph Haydn: Quartet in f, op. 20 no. 5, first movement (exposition)

Musical Example 3.9a continued

Musical Example 3.9b

Joseph Haydn: Quartet in f, op. 20 no. 5, first movement (recapitulation)

Musical Example 3.9b continued

Particular interplays between frame and content, structure and reference substantiated, however, the uniqueness of individual works, made possible precisely by *shared* modes of understanding. Cohesion and flexibility were guaranteed by a cognitively structured world, where musical components presuppose one another, creating an intricate system of expectations, which results in musical narratives of different dynamics and intensities. "Narrativity," writes Treitler, "entails an engagement with time. It is the dimension of the life of consciousness moving in time and organizing constellations of experience as it does so."[53] In music, sonata form treated the overall organization of the experience—highlighting its crucial phases, whereas motivic elaboration revealed how it transpired, capturing the uniqueness of the momentary. It is the latter, paradoxically, which lends permanent nuances to what has happened.[54] The composer's voice notwithstanding, the fictive voices in music are always of an immediate, scenic kind. The "protagonists," in other words, are "on stage," themselves expressing their "states of mind" within the temporal frame of an immediate presence. Perhaps this explains why so many seized on the dramatic metaphor for the description of sonata form.[55]

Though some dramatic elements of the form may relate, in a literal sense, to staged drama, still the narrative seems a more likely next-of-kin, at least as far as instrumental music is concerned. Motivic elaboration bestows upon instrumental works an intensity unwarranted in literal drama. What remains fictive in opera becomes symbolic in instrumental music—heroes, their utterances, their action. These features touch upon the two layered temporal dimension of the narrative—the reading time and the fictive time. The hearing time and the imaginary, metaphorical time of music are not the same either. Lacking reference to literal time, the imaginary time in music refers to a duration outside the ordinary frame of temporal conceptions. It is on this ground that music comes close not only to the kind of experience fostered by a novel, but also takes us back to its metaphysical origin, making room not only for a Kepler and a Bergson but for a Schopenhauer as well.[56] Indeed, transcending ordinary time, music now had the power to express the unfolding of experience in the virtual time, which it symbolized.

Music's begotten illusion, as it turns out, took the communicative power of musical signs for granted, but not what they came to symbolize. The latter, as we have seen, was carefully forged over time.

Paradoxically it is the processes of "acquisition of meaning" that increased the awareness of signs as powerful *cognitive* tools in the making of culture.

Notes

1. *Seconda prattica* was Claudio Monteverdi's term, in the preface to Book V of his madrigals, for differentiating between the old polyphonic style and the new expressive language allocating the latter a cultural place with which he was engaged in that book and in many other collections and works. *Stile concertato*—considered as the forerunner of the baroque concerto, was applied to works consisting of distinct groups of players or singers, small and large, antiphonally or responsorially related. *Stile rappresentativo* served as the generic term for monodic, recitative-like passages, in vocal as well as instrumental music, characterized by irregularity of phrasing, sudden melodic and harmonic turns, freedom of phrasing, frequent pauses and other qualities related to the idea of dramatic "representation."
2. Each of the three main national styles treated expressivity in music in different ways. The Germans adhered to the rhetorical paradigm, and were late in developing their own operatic style; the French were engaged in musical renditions of heightened declamation, rather than in the development of a musical language of the emotions. The French however were also preoccupied with instrumental "exemplifications" of figurative titles. See Bukofzer 1948; Palisca 1981; Katz 1986: 102-107; 185-189; Anthony 1997. The hegemony of the Italian conception—on which we mainly elaborate in the following—does not mean, of course, that the contributions of the Germans, the French and even the English (e.g., the masque and related forms) to this development was not considerable.
3. Monteverdi was not the first to call upon *musica* or one of her sister sirens or muses, to ascend the stage. *Armonia* already appeared in 1589, in Emilio de' Cavalieri's *Dall più alte sfere*—a madrigal from the *Intermedio* "witnessing" *Minerve e Alcide's* wedding (i.e. the wedding of Don Ferdinando Medici and Madona Christiana de Loreno), and *Tragedia* was ushered in by Peri's *Euridice.* Bold muses as they were, they all embarked upon traditional singing, of the kind described below.
4. Parts of Caccini's and Peri's dramas can be interpreted in similar ways, adhering, as they did, to the conception of Greek drama. With Monteverdi, however, the new fictive status of music became more salient and differentiated.
5. Danto, dealing with "Works of Art and Mere Real Things," explains this paradoxical condition as the general enterprise of artistic mimesis: "It is precisely the confidence that the conventions are understood," he writes, "which enables the mimetic artist to carry mimesis to its extreme point, to make whatever is to appear within the relevant brackets as much like what would be encountered in reality as he can manage. And his major problem might be phrased thus: to make whatever is so bracketed sufficiently like reality that spontaneous identification of what is being imitated is assured, the brackets themselves guaranteeing that no one will take the result for reality itself" (Danto 1981: 23).
6. Cone's discussion in his *Composer's Voice* (1974), entails many of the interesting implications this peculiar condition entails. Yet Cone assumes that this state of affairs is essential to music as is, at least to Western music. Arguing that the whole conception was brought to the fore only at the beginning of the seventeenth century,

we do not deny that one did not occasionally hit upon it previously. Undoubtedly, the creators of Medieval Religious Drama and related forms relied on it unwittingly. More important in this respect were some madrigals and religious works of the sixteenth century, such as Josquin's *Absalom*, as observed by Cone. In fact, works of this kind paved the way for the conception that eventually became the major frame of reference. Still, there is a clear difference between the manifestations of this conception prior to 1600 and those that came later. To repeat, there is a major difference between the dynamic force of new developments and the force of their rediscovery. It is only the latter, we claim, that could cause a comprehensive change in the "destiny" of a given domain.

7. Donington drew a line between *Orfeo* and *The Ring* (1963; 1968), attributing to the psychic sphere of the former the rich unconscious activity he imputed to the latter. His basic assumption, however, does not necessarily lead to his specific Jungian interpretations.

8. Such consideration, as we noted, was strongly assisted by the rise of science. Exemplified, as it is, in Kepler's *"Apostrophe"* to *Urania* (the muse of astronomy), it reveals how discoveries in the sciences and the arts seemed to share the same epistemological nature: "Now Urania, a more majestic sound is needed, while through the harmonic ladder of celestial movements I ascend yet higher, to where the true archetype of the world's structure lies hidden. Follow me, modern musicians, and express your opinion by means of your arts, unknown to the ancients. Nature, always generous, has at last, after carrying you in her womb for two thousand years, brought you forth in these last centuries the first true likeness of the universe. By your symphonies of various voices, and whispering through your ears, she has revealed her very self, as she exists deep inside, to the mind of man, the most beloved daughter of the creator" (from Kepler's *Harmonia Mundi* Lib V, chapter vii, as translated in Katz and Dahlhaus 1987 i: 132). Platonic through and through, while exalting the polyphonic music repudiated by the Italians, Kepler's basic position with regard to discoveries was shared by many, including Vincenzo Galilei and his friends, their different musical preferences notwithstanding.

9. Bukofzer 1939-40: 21. Bukofzer's article, which appeared in the *Journal of the Warburg Institute*, approved of the methodology, which dominated the institute at the time and later. Aiming at a "positivistic" iconology and iconography, the scholars of the institute tried to identify objective modes of symbolization, concentrating on artistic styles which seemingly exemplified such objective modes. His contemporary, Walter Benjamin, developed somewhat earlier his theory regarding the allegorical standing of German baroque drama (Benjamin 1998 [1963]). He likewise saw the "hieroglyphic" nature of the baroque semiotic system, but viewed its conventionalism as part of its inclusive nature—a transcendental script *open* to additions, alterations and interpretations. While relevant to music as well, Benjamin's approach does not, of course, treat the unique conditions of music at the period, which were different, as we try to argue.

10. This conception fits in many respects the idea of *impresa* as understood in the sixteenth and seventeenth centuries (Gombrich 1972: 160-165).

11. The antiphonal, echo-like character of this movement and of the work in general, its overriding ideology, and the underlying semiotic structure which supports it are further discussed by HaCohen 1999.

12. He strongly opposed, however, Schweitzer's claim that a continuous development of expressive music existed from Bach to Wagner. Of course, there is a vast differ-

ence between Bach and Wagner (which Schweitzer was the last to overlook), but Bach did not deal exclusively with allegories (as Bukofzer conceived them), nor did Wagner ever exploit naturalistic psychology. Rather, Bach treats the emotional content less as a dramatic process, and more as an emotional state of being. Schweitzer, in fact, was cognizant of both the allegorical as well as the more expressive aspects in Bach's music and, for that matter, in baroque music in general. His mistake was that he confused the two (Schweitzer 1975: 441).

13. See Schweitzer 1975: 51. Schweitzer's many examples are dispersed throughout the book, but his musical "archetypes" are mainly explicated in the pages following the above quotation. From a symbolic-musical point of view, they lack systematic order, nor do they stand an "empirical" test, when checked against Bach's *oeuvres.* Still, this should not weaken the argument that, if indeed musical passions are organized in binary relations, then sub-emotions relate to "shades" rather than to a "multiplicity" of distinctions and so does musical coinage, as will become clear later.

14. Centering on the so called "common" meter and on the 3/4 triple meter, such a "continuum" comprises, at the High Baroque, the following spectra:

Ecclesiastical Galant
(exalted passions) (terrestrial passions)
4/2 C 2/4 4/4 3/4 6/8 3/8 9/16 9/32

Changes in the "scale" were introduced later in the century, when the number of metrical alternatives were reduced, and, at the same time, the metrical flexibility within a given "number" increased (Alanbrook 1983: 1-70).

15. Cooke, however, ignores the historicity of these correlations, arguing, like Rameau, that the major/minor harmonic system exploited the tonal tensions inherent in the basic row of the overtones.

16. Speaking of the full "acquisition of meaning" by the eighteenth century does not imply that early seventeenth-century composers had to invent a musical vocabulary from scratch. A glance at Monteverdi's *Orfeo* will suffice to get an idea of the existing coinage which he so ingeniously exploited: The joyous frottola with its hemiolic rhythm to express Orfeo's joy before his fatal turn; the pastoral ditties, interwoven into the pastoral scene of the wedding; the quasi-polyphonic choral pieces for the reflective moments of the drama (Compare this vocabulary to the above description [p. 35], following Hollander [1957], of the musical coinage alluded to in Shakespeare's *Twelfth Night*). Likewise, the minor mode was already associated with tragedy and grief; the effect, otherwise, of the changing tonality upon the entrance of the *messagiera* in the middle of Act II would have simply been meaningless. As is well known, the affective import of the minor triad was already felt by Zarlino; Kepler tried to explain the phenomenon "scientifically" (Katz and Dahlhaus 1989 i: 117-20).

17. Musical allegories of this kind intellectually resemble the pictorial allegories of poets like Pope and Thomson, discussed in chapter 2.

18. To emphasize our point, the reader is asked to imagine, for example, Caccini's *Euridice* in an instrumental rendition, like those so popular with Mozart's operas.

19. Glarean's inclusion of these modes in his codification of the modes was a recognition of a musical practice already in existence.

20. For the beginnings of thorough bass in practice and theory see Arnold's classical study (1931/1965 i: ch. I). See also Lester's (1992) recent study.

21. Translation: From the magic cavern, strident hinges, open an outlet for me/ and through the darkness of the black shelter let me in!/ high on the horrible alter of the Stygian waters the flames are glittering/ and upward climbs the smoke that whirls the light to the sun.
22. For the central position of melody types in the Gregorian chant see Apel 1958.
23. These developments exemplify what Weber (1958) called "the rational foundations of music in the West." As far as development of harmony is concerned Bloch seems to ascribe some of these achievements to composers as early as Lassus, for he writes: "from this time on only chordal hearing, that is, a unifying that is vertical below and horizontal only above, where the overall impression of the melodic element occurs, can be reckoned an adequate means of grasping these works" (Bloch 1985: 5).
24. Narmour (1992), among others dealing with music cognition, leans in this direction in his attempt to explain the perception of music. See also Watkins & Dyson 1985.
25. This process is described by Wortshorn 1954.
26. Once this "grammatical" condition had crystallized, it gave rise to a rich variety of interactions between text and music on the spectrum between coloratura (relying on motivic elaboration) and syllabic singing. This enhanced the musical repertory of dramatic situations and expressive modes as demonstrated in Reinhard Strohm's studies of early eighteenth-century *Drama per Musica* (Strohm 1997).
27. Addison and St. Evermond are known for their harsh criticism of Italian opera. Important excerpts of their writings appear, respectively, in Strunk 1981 iii: 151-7; Weisstein 1969: 30-35.
28. For an insightful analysis of the interrelationship between vocal music and instrumental music at the period see Georgiades 1982: 66-83.
29. In Bach's *St. Matthew Passion*, in which almost all of generic and symbolic modes related to baroque music are *fused* together, the *da capo* principle governs not only arias but the huge choruses as well. In the "real" arias of the second part, however, it is embodied in a "through composed" way, resulting in an intensive arch form; the *cori spezzati* principle, "imported" to Germany by Schütz a hundred years earlier, build the dramatic force of the *turba* parts. Combining different procedures (e.g., arioso and choral) within one and the same "number," it fabricated a reality of a more complex nature than the one rendered in the opera.
30. The division of the whole polyphonic setting into two "antiphonal" parts that are always related to the bar line became a central textural property of the new style.
31. About the predominance of the synthetic over the analytic as perceived by listeners, even in the most "semantic" of all musical utterances—Richard Wagner's leitmotivs, see HaCohen and Wagner 1997.
32. Different from all the rest, even from Johann Scheibe (who sought to explain the change via the *Figurenlehre*) was Rousseau. Though he could hardly explain the underlying mechanism of the "scaffolding system" he seems to have understood its nature: "The symphony itself," writes Rousseau, the admirer of Leo, Pergolesi and Jomelli, " has learned to speak without the help of words, and often the feelings that come from the orchestra are not less lively than those that come from the mouths of the actors" (from a letter to d'Alembert, 1754, as quoted in Le Hurrary & Day 1981: 91; see also HaCohen 1998).
33. Meyer (1980) rightly conceives the classical "period" as an archetypal structure, which later functioned as a frame of reference for intentional deviations aiming at certain expressive or stylistic statements. Later, together with Rosner, he tried to support his conception empirically (Rosner & Meyer 1982).

34. Wolf was aware of the historical derivation of Stamitz's style. However, it is the elaborate scaffolding grammar, which the Italians developed, that enabled Stamitz to develop a variety of musical phrases suited to various structural functions.
35. The fact that Cone (1974) uses Berlioz to exemplify his conception of the role of instrumental parts as "virtual agents" is, of course, not accidental. With the romantic composers, instrumentation achieved full fruition, granting the instruments the richest dramatic, expressive and atmospheric roles they had ever enjoyed. As we argued elsewhere, looking at the roots of such qualities discloses the semiotic web in which they were posited from the start, and the cognitive implications they harbored.
36. "The composer's persona," according to Cone, is a compound arising from a mixture of the vocal and the instrumental, or "by the unifying power of the musical line." Cone warns us that this does not mean "*the* persona of the composer," but "*a* persona of the composer"; for the persona of each composition is "uniquely created by and for that composition" (Cone 1974: 18).
37. This statement follows Abrams (1953) well-known claim concerning Romantic poets. One may argue that it is the control of different kinds of musical utterances, their degree of saliency and their expressive quality, that enabled the classic-romantic composers to conceive their works in this way.
38. For example, C.P.E. Bach's sonata forms are motivically oriented; those of Stamitz strive towards compactness and fluency, whereas those of the early Viennese symphonists are of contrapuntal interest. See Barford 1965, Newman 1963, Churgin 1963, Rosen 1980.
39. In this connection, Kerman warns us not to confuse the metaphorical with the literal (Kerman 1952: 76-77). Rosen tells us that "it was, in fact, under the weight of this dramatic articulation that the High Baroque aesthetic collapsed," (1976: 43).
40. Rosen 1976: 26-27; also see Kerman 1952: 75.
41. On the whole, much of the inner conflict of the form is based on the interaction between the period on the one hand, and the vast tonal areas on the other. The fast tonal rhythm of the development thus reflects the breakdown of the period, enabling through fragmentation, construction and a further deconstruction the building of the whole development section on the "natural" passage of growth, climax and denouement.
42. Treitler 1988: 421. Kerman also notes that the recapitulation ordinarily contains "some new reflection of some prior detail," which "seems to display in small the new ascendancy over the earlier conflict" (Kerman 1952: 79) yet he was not aware of the narrative nature of the phenomenon.
43. The transition from a thematically oriented theory of plot to a functional theory was crucial; Propp, who introduced this revolutionary idea, considered function as "an act of dramatic personae, which is defined from the point of view of its significance for the course of action of the tale as a whole" (Propp 1968: 20) Later narratologists, from Levi-Strauss (1963) through Barthes (1974), Greimas (1977), Todorov (1977), Bremond (1970) and others, adhered to this conception. This is also prevalent in cognitive studies of narrative, which aim to disclose its grammar on a more empirical basis (Rumelhart 1975; Shen 1989; to mention but a few).
44. Summarizing the main claims of the narratologists mentioned above, Culler claims that "those aspects of the movement, from the initial situation to the final situation, which help to produce a contrast between a problem and its resolution are the components of the plot" (Culler 1975: 213).
45. Recent attempts to analyze the understanding of narratives in terms of a special cognitive skill or as related to a meta-cognitive grammar (e.g., the x-bar grammar;

Shen 1989) lends support to the view that narrative can be manifested in different media, even in a medium as a-lexical as music. However, even if this is possible, it should be viewed as a metaphoric manifestation, as we shall immediately argue.

46. In his study of the late symphonies of Haydn, Schroeder (1990) endeavored to show how Enlightened thought, and particularly ideas concerning morality, affected Haydn's music-making at the time. Haydn, we learn, wished to "translate" some of these extramusical thoughts into a musical unfolding that "would lead his audience through a process of discovery, the importance and function of duality, and the arrival at an understanding of tolerance" (p. 20). These moral-philosophical ideas draw upon Shaftesbury's view of the role of art and the artists. Combining moral philosophy and aesthetics, Shaftesbury maintained that the artist might and should create a context that allows for the coexistence of opposing forces. Based on persuasive contemporary evidence, Schroeder shows how Haydn came under the spell of these ideas, aiming to convey them musically, and believing himself to have succeeded in "presenting a musical demonstration in instrumental music of the idea of tolerance" (ibid.) Schroeder also endeavored to show how the Enlightened audience understood this musical "translation" (pp. 95-122).
47. See Powers' (1980) criticism of Strauss's comparison of music and myth.
48. This becomes clear from many recent studies that deal with narrativity in nineteenth century music and beyond. See for example Newcomb 1984; 1987; See also Hatten 1994, especially chapters I and III.
49. Novels built on the precepts of sonata form are discussed in Brown 1948; Aronson 1980; Basilius 1981.
50. The Viennese were also those who criticized the earlier achievements of the classical style—especially of the Mannheim composers—for being rhetorical exercises void of content. Leopold Mozart, for example, commented that their symphonies "are hodge-podge, with here and there a good idea, but introduced very awkwardly and in quite the wrong place" (King 1955). It is the insufficiency of the galant style that prompted people like L. Mozart or Johann George Sulzer to warn the composer not to be attracted by "the pleasant so-called *galant* pieces, and thereby introduce small, chopped-up, dainty music instead of beautiful music into serious works" (Ratner 1956: 440). Kirnberger expressed similar views; For the meaning and extension of the Galant style see Sheldon 1975.
51. The Viennese style and the penetration of the contrapuntal writing to the prevalent style is discussed by Larsen 1967; Webster 1974; Kirkendale 1979i.
52. In the best cases the two (energy and physiognomy) are combined. See, for example, how in the first movement of Mozart's g minor quintet (K. 516) a special combination of g minor ascending arpeggio and its chromatic reversal, scaffoldingly structured in an enlarged period, makes for and extended "first theme" area. As such it compels an elongated struggle for the implied musical agency, to get free of its tragic fetters. This is embodied in a new figuration of the first motif (highlighting the sixth of its sixth-fourth permutation) which carries this dramatic development through intervalic extension.
53. Treitler 1988: 420. In his insightful essay on "Music and Duration," Basil De Selincourt discusses music as a form of duration which suspends ordinary time, and offers itself as an ideal substitute and equivalent. Our life and consciousness, he argues, are more closely conditioned by time than by space. Music demands "the absorption of the whole of our time-consciousness; our own continuity must be lost in that of the sound to which we listen... Music uses time as an element of expression; duration is its essence" (see Langer 1961: 53).

54. See Gabriel Marcel's excellent essay on "Bergsonism and Music" (see Langer 1961).
55. Abbate (1991) is most explicit on this point, but her argument does not take into consideration the complex metaphorical status and qualities of musical narrativity, stemming from the medium and its basic constrains, as explicated in the following paragraph.
56. Kepler was the first to understand the metaphysical potential implied in a musical narrative. He writes: "By the artificial symphony of several voices he [modern man] plays out in a brief portion of an hour the perpetuity of the whole duration of the world, and thus tastes in some degreee the Creator's satisfaction in his own works, with a most intensely sweet pleasure, gained from this music that imitates God" (Katz and Dahlhaus 1987 i: 139).

4

Signs as Cognitive Tools

The development of modes of signification in Western music, involved, as we have seen, semiotic procedures familiar to us from other communicative systems. Powerful tropes such as allegory, metonymy, and metaphor were musically configurated; interrelated referential schemes, syntactic structures, and narrative frames became tonally encoded, and entrenched symbols and flexible imagery transfigured each other musically. The cohesion of these schemes and procedures transpired in an illusive level, revealing their efficacy in the fictive world that they fabricate. Tough some of these schemes and procedures were surveyed in the *trivium*, they were not, however, discussed in cognitive terms nor from a shared semiotic point of view. Interestingly enough, while these musical developments took place, a parallel development took place in philosophy and literary thought, preparing the ground for later thinkers to deal with artistic phenomena and their illusive nature. This entailed, first and foremost, an investigation of signs—cultural rather than natural signs—as cognitive tools, in search of the mental processes they instigate, exploring, as well, their role in the making of cultures including that of the art.

Signs and Symbols Redefined

The communicative power of signs has been subject of persistent curiosity since the early history of culture. In most ancient cultures the control of signs was entrusted to the select few, treated as though belonging to a super-natural order. The Greek philosophers, who tried to analyze the epistemic efficacy of signs, contributed toward their demystification. The stoics notwithstanding, the Greek generally held that signs could be defined and thus substituted by a "semantic synonym."[1] Their epistemology reflects the same logic; idea, they maintained, is mainly located in the mind whereas the correspondent object in the world of *realia*. Whereas idea meant form, and mind the mirror in which it is reflected, object was basically identified with the assumed form. Despite his empiricist outlook, Aristotle supported this conception, holding that form, associated with a certain object, is inflected through the properties of a certain medium (e.g., air) to the sense organ and the mind.[2]

These Aristotelian assumptions were later integrated within neo-Platonism, and as such kept alive into the Renaissance, despite occult practices. Yet occult traditions insisted on the "magical" power of signs, not only to represent but *create* desired situations. Ficino's music-spirit theory (discussed in chapter 1) attests to both trends.[3] Interesting in this connection are various symbolic devices—hieroglyphs, emblems, imprese, ensigns etc.—which entered into theories of symbolism in the fifteenth century, when interest in such traditions rose (Gombrich 1972). From the fifteenth to the sixteenth century the uses of these symbolic devices underwent refinement through various kinds of codifications, which resulted in learned compendia available to those who wished to represent matters in a proper symbolic fashion. Through signs understandable only to the initiated, this tradition was largely applied to visual experiences, penetrating, as such, poetic thought.

What began as a unity of the symbol and the symbolized, led in due course to recognition of the artificiality of the one and the dependence of the other on inculcation. Artificiality, however, does not mean arbitrariness, since the choices depend on the communicative potential of the symbol, as explicated by Benjamin (1998). That the lion became the king of animals is no coincidence, that the tiger did not, reflects a choice. The lion, however, may stand for a variety of qualities—strength, vigor, vigilance, etc.—leaving a margin of ambiguity, inviting creative encoding.

Declining though they were in the seventeenth century, occult traditions continued nonetheless to play a significant role in the attempts to explicate the power of symbols in the creation of worlds of knowledge, including the arts. Of course, bona fide science, as we have seen in the case of Fludd, had no use for such symbols. Ironically, however, an attempt such as Descartes's to deal with the passions of the soul, from the point of view of its symptoms as well, led to LeBrun's compendium (1698) that was conceived in the spirit of the books of emblems. In England, the emblematic tradition continued to flourish all through the seventeenth century (the expression of which we have encountered in the picturesque poetry as well as poetry, which availed itself of musical imagery). Yet, from Bacon onwards, the tendency to demystify symbols characterizes the British. By the end of the seventeenth century, these trends of thought spilled over into the philosophical deliberations which coped directly with the epistemological implications of the new scientific methods.

These philosophical deliberations also concerned themselves with the role of signs, as it emerged from the dichotomous notion of idea and object. It was this dichotomy that disturbed Descartes. For Descartes, essences were no longer forms perceived but proportions conceived, establishing mathematics as the core of epistemology. Reality was thus split into two kinds of qualities; the primary—clear and distinct—were considered the product of rational deliberations, whereas the secondary—clear and confused—were conceived as the sensual effects of the primary qualities.[4] The mind was thus thought of as a bundle of faculties, each of which responsible for the "translation" of some kind of mathematical data into sensual formation of sorts—color, sound, and so on. This kind of "innatism" implied that sensual ideas could be formed even without external objects; this, as we have seen, holds true for Descartes's theory of the passions as well.

Primary and secondary qualities became an underlying premise for most of the believers in the new science. Yet for those who did not expect God to guarantee their epistemology, mathematics was considered but a methodological tool improving partial and subjective empirical findings (Cassirer 1951: 96-7), as we have witnessed in the case of Mersenne. Locke was unequivocal in this respect, claiming not only that epistemological considerations start from secondary qualities, but that the mind itself works in that order, arriving at primary qualities through inferences. Process—in the Cartesian sense—was thus turned upside down, yet the problem of perception rose again. Interestingly enough, Locke returned at this point to conceive "idea" as the representation and the perceptual act intertwined. Perceptions, he thought, are impressed on the mind's *tabula rasa,* becoming "the object of the understanding of a man thinking."[5] In the cognitive process, he assumed, developmental stages, rather than mechanical components, played the main role. The mind's power, consequently, became itself an object for introspection; learning abilities thus rose in importance, weakening the theoretical necessity for more innate and defined dispositions. Locke seems to have connected, perhaps for the first time, semantic inquiry with the theory of knowledge (Kretzmann 1977).

Locke's "ideas" suffered nonetheless from latent innatism. For him, simple ideas were already distinguished by Cartesian distinctness, which enabled their immediate enlisting as components of rational thought. Small wonder that he viewed language as a substitution of linguistic labels for preexisting mental entities. Likewise, he considered propositions as nominations for relations already fixed by prelinguistic thought (Kretzmann 1977: 126-128). In this way he could account for the conventional and arbitrary nature of natural languages, while keeping the old conception of language almost intact (Yolton 1985: 120). According to Locke, language comprises a mirror of a natural, primitive yet reasonable mind, itself reflecting a reality rationally constituted. Free associations, let alone imaginative deviations, he therefore considered as erroneous activities of the mind (ibid. 135). Such ideas lent support to the hegemony of scientific language in seventeenth-century England, penetrating poetry as well, as we have seen in chapter 2.

The problems inherent in Locke's theory of knowledge are reflected in the way he conceived the operations of the mind. Impressions, Locke maintained, become ideas only through their conscious isolation; what is learned is thus explicitly known, acquired through processes that the mind is *aware* of. Reflection, hence, consists of the mind's ability to trace its own activities and operations.

Ironically, it was Leibniz, the rationalist, who warned against Locke's too strict account of the mind. Substituting the Lockian *tabula rasa* with the metaphor of a "block of marble with veins," he could entertain the thought that perceptions are neither logical nor distinct (Harris 1977). In fact, he observed blur and confusion in perception, which he attributed to the *lower* cognitive faculty. Unconscious calculations, which Locke rejected, played a central role, for Leibniz, in the transition from the unlogicized to the logicized constructions of thought.[6] Leibniz, thus, emphasized innate *operations,* rather than innate ideas, a view that consequently led to an implicit separation between innate ideas and universals.

However, it was the "higher cognitive faculty"—related to logic and science—to which Leibniz paid his main attention. Believing, as he did, in preestablished harmony between mind and reality, he avoided the question of the emergence of primary components of thought, and, like Locke, entertained the notion of simple ideas, which he conceived, however, as *artificial* components of thought. The mind, endowed with its own powers, sifts these components in order to further their skillful manipulation. No wonder that Leibniz was able to realize, more clearly than others, that the nature of such mental components is symbolic. Explicating their cognitive efficacy, he tells us that

> no one should fear that the contemplation of signs will lead us away from the things in themselves; on the contrary, it leads us into the interior of things. We often have confused notions today because the signs are badly arranged, but then with the aid of signs we will easily have the most distinct notions, for we will have at hand a mechanical thread of meditation, as it were, with whose aid we can easily resolve any idea whatever into those of which it is composed.[7]

Signs, thus, function as focal points of reasoning, guaranteeing order, clarity, and completeness. While they depend, as such, on the intension or content of that which they signify, they set up at one and the same time syntactical structures, which are open to redevelopment. It was a well-made system of this kind which Leibniz was after in his search for a mathematical calculus which could approximate the "real design"—synchronic and diachronic—of the "universal plenum." He thus arrived at his infinitesimal calculus with the aid of which he was able to exemplify how certain concepts are vested in others to the effect that the formers generate the latter. Indeed, the differential-integral calculus is, perhaps, the first and one of the finest examples of a generative system in which basic "germs" contain the very rules that enable their further transformations.

Well-defined as Leibniz's signs were, he nevertheless continued to perpetuate the old notion of "synonymous semantics." Yet fully aware of the artificiality of his signs, Leibniz was soon to realize their constructive power, which extends beyond the given reality. Through signs, guaranteed by logical and scientific reasoning, the mind, Leibniz maintained, is free to create its own possible worlds, independent of ontological constraints. Though signs still stand for their definitions, they might enlarge their basic syntax, if not the nature of the signs themselves, through an extension of the given.

Leibniz also went beyond "synonymous semantics" in his new numerical system—the dyadic system, which entails two abstract components—"One" and "Zero." Not accepting the traditional identification of Zero with *nihilum* he defined Zero as the "confining factor" for every real number.[8] The different combinations of binaric pairs became for Leibniz a more accurate expression for the unification of manifold substances into unique and independent wholes—the monads. In the organic world, which the monads populate, each comprises its own special structure, representing the whole network in which each plays its special part. In other words, "monadology" calls for holistic semantics, in which each monad is circumscribed by all the others through endless chains, as Eco would argue (Eco 1984).

Interesting and far-reaching as Leibniz's ideas were for future development,[9] they only dealt with signs of a purely conceptual nature. Envisaging a language that would not only function in accordance with such a conceptual scheme, but that would also encourage thought and creativity, Leibniz, like Mersenne, confused the conceptual with the perceptual and the emotional. Inspired perhaps by Mersenne's writings, he even enlisted music for that purpose, trying to disambiguate language through a musical notation of sorts. Erroneously, he expected this "conceptual music" to evolve into "beautiful and moving songs," believing, in a true Platonic spirit, that the beautiful, the true and the good are but different aspects of one perfect order.[10]

Leibiniz's confusion of the perceptual with the conceptual discloses nonetheless a bent towards empiricist ideas as they were taking shape at the beginning of the eighteenth century. Like the empiricists, Leibniz viewed the direction of "cognitive" processes as "down-up" processes that exploit sensual components as their basic materials. Like them, he saw the mind as characterized by potential operations which it must learn to activate. Moreover, starting with basic perceptions and ending up with the most abstract conceptions, these processes, Leibniz believed, enable us to differentiate between external and internal, between the world and ourselves as conscious beings. In our basic perception we can, indeed, isolate the sensational components from their organization in space and time. The acknowledgement of the sensual traces in the most abstract intellectual activities contributed to an awareness of the concrete aspects which symbols entail.

Whereas Locke's theory of mind concerns primarily propositional thought, Leibniz's philosophy leads again to symbolic systems the nature of which is less clear and concrete. It is no coincidence therefore, that Leibniz should find interest in an English contemporary—whom we shall yet discuss at length—the Earl of Shaftesbury. The point of departure for both Shaftesbury and Leibniz was the harmony of the plenum, which found a new expression through science. However, whereas Leibniz tried to find in this *"Reimung der Dichtung"* a mathematical equivalent by means of his infinite calculus (not unlike Kepler), Shaftesbury was after the secrets of its beauty (not unlike Fludd). Drawing as they both did on Platonic sources, Leibniz's orientation was, in the main, Pythagorean, whereas Shaftesbury's, as a disciple of the Cambridge Platonists, was, in the main, Plotinian. The old identity of truth, beauty, and the good occupied Shaftesbury in his attempt to find a solution to the problem of the truth-value of sensation.

"All beauty," says Cassirer in the name of Shaftesbury, "springs from truth, but the full, concrete meaning of truth can only be revealed in beauty." Indeed, Shaftesbury was not concerned with a "sum total of theoretical knowledge," but, rather, with an *inner* structure of the universe, which can be known only through intuitive knowledge. This should not be linked with the psychological aspects of experiencing art, since intuitive knowledge of beauty reveals itself primarily, according to Shaftesbury, in the creative act itself, in the *partaking* in the knowledge which God entertained in the act of creation. Thus, through artistic acts, one may encompass an all-inclusive truth. Shaftesbury's view might be interpreted as an invitation to artists to explore "possible worlds," worlds hitherto unknown. One can readily see how an orientation of this kind was more likely to lead to an exploration of "the art symbol" as a whole, rather than of the symbolic components within it.[11]

Indeed, though Shaftesbury was vehemently against hieroglyphics, he saw the art of painting functioning like an emblem. He may, perhaps, be considered the first to lend painting an independence from verbal descriptions, an independence based on an organic unity paralleling nature itself. Shaftesbury stressed the necessary compatibility between the organic design and that which it tries to convey.[12]

These ideas appear in the context of his threefold classification of "writings." The first, he calls "notes," is a kind of shorthand writing. The second, which he calls "signs," consists of imitations. The third, which he refers to as "mixed" results from a combination of the former two. The writing which indicates phonetic entities, those identified with "notes," are irrelevant to the language of art; they indicate the arbitrariness in natural languages, which may be prejudiciously manipulated. Only the other kinds of writing pertain to the "fine, beautiful, and just," for they take into account "natural history, passion, habit, and form" (Dieckmann 1970: 117). Of interest is Shaftesbury's implied suggestion that the development of mankind may be traced through these various modes of communication, brought about in the course of creating civilizations. It seems as though Shaftesbury's organic aesthetics spills over into a concept of organic growth, creating a new perspective with regard to the whole notion of hieroglyphics. Noteworthy is also the position he grants art, placing the aesthetic on top of a pyramid, which takes into account the accumulation of culture.

The Cognitive Nature of Signs

The link between the development of mankind and the various modes of communication, which is merely implied in Shaftesbury, is fully explicated in Bishop Warburton's *The Divine Legation of Moses Demonstrated from the Principles of a Deist* (1738-41). Less interested in the aesthetic and more in establishing the foundations of a "natural religion," he aimed at demystifying both the Old and the New Testaments. Accordingly, he created a theory whereby stages in the development of writing are correlated with the development of types of discourse. Writing entails three stages—the first, that of pictures (in which painting directly conveys that which one intends to express), the second, that of hieroglyphs (in which analogous and "symbolic" means are employed) and the third, that of an "alphabet" (consisting of conventional, abstract signs). Whereas the first two stages relate signs to things, the third commences with signs referring to whole words and ends with signs indicating single sounds.

The three kinds of discourse to which they are respectively related (theoretically, though not necessarily historically) are language of action (gestures and movements which help to convey meaning), language of fables (abounding with elaborated similes) and metaphorical language (conveying sensuously abstract ideas).

Warburton insists on the interdependence of the kind of symbol employed and the stage in the development of culture which gives rise to it. Hence, he argues that modes of discourse can be understood only within the cognitive constraints imposed by the stage to which they belong. Thus, for example, he interprets biblical "supernatural visions" as a kind of poetic discourse that employs a mode of conversation abounding in action. "The *delighting in unusual Actions and foreign Modes of Speech, is an Indication of the Turn of Mind*," which is informed by action, tells us Warburton. "When it is shown that those (expressions) in question are idiomatic and familiar, the Suspicion must drop of course.... When the *Sacred Writers* talk of being *born after the spirit*, of being *fed with the sincere Milk of the word...* they speak the common, yet proper and pertinent Phraseology of their Country; and not the least Imputation of *Fanaticism* can stick upon those original expressions" (from Warbarton's op. cit., vol. ii, p. 86, quoted in Barry 1987: 7). Unto itself, the "language of action" still points, for Warburton, to an inadequacy in representation. Indeed, such signs point either beyond themselves or convey less than is expected of them; these "drawbacks," however, will yet turn to be their very advantages, calling attention to processes of interpretation.

Strengthening the British rational approach toward hieroglyphics (from Bacon to Addison), Warburton, nevertheless, points to "mystery" as an important component in the development that he delineates. Maintaining that the symbolic stage, in both writing and discourse, made earlier signs mysterious, he implies that symbols stand for more than "meets the eye," that they entail their own historicity. It is no wonder, therefore, that for Warburton the developmental stage that coincides with the alphabet is the highest mode of discourse: unlike Shaftesbury's "notes," it incorporates all stages in cultural development, including the iconic stage. Whether aware of the prerequisite cognitive development that his abstract stage entails or not, the fact remains that Warburton distanced himself from Locke's view of language as a natural reflection of a given, rational consciousness.

Altogether Warburton's three stages testify, Barry tells us (Barry 1987: 7), to a "principle of economy, by which each of these forms replaces its predecessors, and by which the world can be represented in a more abbreviated and efficient manner." Still, Warburton insisted that the various types of language—language of action, similes, and metaphors—were preserved in the later forms of speech. Warburton's ideas, including the last one, as we shall see, were echoed in Condillac's elaborate explication of the development of language.[13]

It was left to later thinkers to further explicate the transitions from the sensual to the perceptual and from the perceptual to the linguistic and the rational, which the thinkers we have discussed so far may have initiated but did not themselves settle. Berkeley and Hume, who continued the epistemological debates of Descartes to Leibniz, were interested in images, not as they relate to hieroglyphs and emblems, but rather in their functioning in processes of understanding. Abstract ideas, they maintained, are particulars which become general through their representational function; both contended that in conceiving an abstract idea, "the image in the mind is only that of a particular object, though the application of it in our reasoning be the same as if it were universal." Hume added that "a particular idea becomes general by being annexed to a general term" that is, to a term, which from "a customary conjunction has a relation to many other particular ideas, and readily recalls them in the imagination."[14] It is, thus, the representative aspect of symbols that makes possible not only reasoning but the imaginative activation of the mind, as well.

The distinction between the sensational and the perceptual and the transition from the one to the other was only vaguely understood. Hume was perhaps the first to recognize this vagueness by distinguishing "impressions" from "ideas": "Impressions," for Hume, meant perceptions experienced in "real" time, hence of perishing existence, "ideas," by comparison, are "less lively perceptions, of which we are conscious when we reflect on any of those sensations." Ideas, therefore, are of a more lasting nature and may be redirected toward impressions as well as associations and relations, which are the mind's imaginative production. The imagination itself, however, is constrained by experience and habit. "Experience is a habit which instructs me in the several conjunction of objects from the past," Hume tells us. "Habit is another principle, which determines me to expect the same for the future; and both of them conspiring to operate upon the imagination, make me form certain ideas in a more intense and lively manner, than others, which are not attended with the same advantage" (Yolton 1985: 179). Interesting as Hume's theory is, it should be noted that Hume still viewed the transition from the sensational to the perceptual as immediate, and though admitting its unlogical nature, he still considered the transition from the perceptual to the prelinguistic mainly in mechanical terms of force and duration. Hume, thus, irrationlized Locke's copy theory of ideas without, however, changing its basic nature.

"Irrationalizing" the mind, was, however, of crucial importance. Believing that all complex ideas relate to the mind's basic modes of association—similarity, causality, space and time—rather than to a reality out there, Hume deepened the gulf between reality and the mind. After all, it is the principle of imagination "which makes us reason from causes and effects; and 'tis the same principle, which convinces us of the continu'd existence of external objects, when absent from the senses" (Yolton 1985: 179). Like his contemporary Baumgarten, Hume relegated the combinatorical power of the mind to that which Baumgarten called the "lower cognitive faculty." We shall see that Baumgarten also considered similar modes of association as crucial to the activation of this faculty. However, for Hume, unlike Baumgarten and his Rationalist mentors, no higher faculty stood above these modes of thought, since he saw even scientific thought, based as it is on causal relations, as conveying no more than customary links which prove efficacious.

According to Hume, thus, science and art share a common epistemological ground, though a totally different one from that held by seventeenth-century thinkers. Hume seems to have secularized Leibniz's view, doing away with the latter's "best of all possible worlds." The difference between the British and the Germans notwithstanding, both conceptions shared recognition of the autonomy of the mind and its constructive powers. It was upon this common denominator that a fertile dialogue between them—especially concerning aesthetics—could have evolved.

However it was a Frenchman—Etienne Bonnot de Condillac (1715-1780)—who perhaps made the most important contribution to the understanding of the cognitive nature of signs. Writing his

Essai sur l'origine des connaissances humaines (1746) as "A Supplement to Locke's Essay on Human Understanding," he tried to develop a tighter theory of cognition. His main thrust was to show that it is only through utilizing signs that the mind learns to control thinking, even in its most primordial stages. His theory entails stages of development in which each stage is determined by a preceding one, but its realization is not ensured thereby. The decisive stage at which the mind begins to use signs at its own will, accordingly, may not be shared by all creatures; what separates "brutes" from human beings lies at the threshold of this stage. "Perception" (Condillac's term for the mind's reception of the sense data) leads to "consciousness"—a mere awareness of perception; the mind, Condillac tells us, learns to concentrate on certain perceptions over others through "attention." Directed by wants and desires, this course of development further leads to "reminiscence," that is, to the recognition that certain perceptions did occur. Four different cognitive stages are thus needed for the mind to become fully aware of that which it perceived. Condillac thereby bridged Locke's postulate concerning the conscious nature of mental activities with Leibniz's contention regarding unconscious processes, rendering consciousness as part and parcel of cognitive growth.

It is only when pressing needs and demands arise that the mind activates "imagination," reviving images through recollection. In "contemplation," awareness is enlarged so as to include the circumstances and the "name" of that which has vanished out of sight, that is, the whole context in which it appeared. Finally, this complex image is retained in memory and connects with similar images through a "*liaison des idées*." Altogether, while the first four stages consist of the ability to individuate perceptual units by keeping them in mind, it is only from the fifth stage that these separate units connect with other such units to create a new reality.

These separated units are Condillac's primordial signs. Condillac unequivocally emphasized that it is only the use of signs, which puts the mind in command of itself, being no longer a merely passive receiver. Through signs "the mind can recall ideas at will, compound and decompound them, distinguish and abstract and compare, affirm and deny, and ultimately reason. It is only by the power of signs that mere consciousness can pass over into meaningful reflection, and the mind becomes aware of its true strengths as well as to its limits," summarizes Stam (1974: 4).

Evolved from associations with specific objects, whether accidentally or otherwise, the above-described signs are all natural and their activation is mainly instinctive. Such signs, disconnected and sporadic as they are, Condillac soon realized, cannot account for the formation of a systematic language. For signs to evolve into language, they must be of a more compound nature. He thus entertained Mersenne's *Gedankenexperiment* of two isolated children, developing a mode of communication with each other. Assuming that compassion motivates human interaction no less than desires, he speculated that they would invent a "language of action." Such a language would be composed of natural gestural signs—single motions or ejaculations clinching whole representations. By gestures of this kind, he further speculated, the children would succeed in conjuring up interrogations and other "modes of information." The more the imagination is enlisted the more durable the impressions that would be left behind. Over a span of time, specific gestures or cries would come to be associated with particular causes of concern, sources of contentment and the like. Interestingly, it is passion, evoking sympathy, that transforms the natural into a social phenomenon. Passion, thus, may turn from cause to referent, with its own characteristic signs—the particular inflexions of the voice. It is this kind of differentiated signs, interacting with other such signs, that enables the development of a linguistic system.[15] In contradistinction to accidental signs, gestures, as Rousseau was soon to maintain, do not derive from outer perceptions or any other preexisting model. While identifying gestures with mimic and intonation, Condillac emphasized their iconic character, insisting, moreover, that gestures, though unanalyzable in terms of their componential parts, constitute complete utterances, a basic condition of language.[16] Language, for Condillac, includes contextual surroundings, which render perception persuasive. To achieve a linguistic system, the "language of action" had to be decomposed, whereas the relations among the components obtained had to be conceptualized through syntaxes of sorts. Signs institutionalized into systems not only permit economy and convenience, variety and subtlety, but they also allow the formation of new combinations of ideas through the juxtaposition of names, verbs and propositional terms; in short, they further our knowledge of the world. Civilization begins when the natural is socialized, when "accidental" and "expressive" signs are institutionalized. Institutionalizing signs, however, also results in some cognitive loss, hence the importance of "languages of action" beyond their primitive role. Indeed, such languages, maintains Condillac, are employed in the artistic domain.[17]

Indeed, music looms large in Condillac, since it performs well the function of communicating emotional states. This, in fact, was the reason for his attempt to trace the early stages of language through the emotional powers ascribed to music. Since there is room *at all times* for messages which are not crystallized, in which neither the sign nor the signified is clearly defined, "it must have appeared as ridiculous to the Ancients to separate the music from words as it would be for us to separate the sound of our theatrical recitation from verse," maintained Condillac. Music, accordingly, functioned metaphorically from its very inception, remedying a "gap in the vocabulary," as Black would have it. Rhetoric, as it were, is embedded in music, and it is from early prosody that the different arts emerged, according to Condillac (Condillac 1974: 208-227).

However, while metaphors of this kind became increasingly dispensable with the development of language, the identification of different configurations of sound with meanings of sorts, lent music a standing of its own. The separation of language and music was made possible, tells us Condillac, because their common origin entailed communication as well as expression. If instrumental music possesses certain emotional meanings, Condillac tells us, the reason is anchored in their common "linguistic" past. Whereas man came to use language to communicate in ordinary life, he "came to hear music with the most favorable disposition" (Condillac 1974: 212).

The "golden age" of music did not, however, last long. Music as an autonomous system of signs, Condillac seems to imply, is doomed to fail, for it invariably needs a "lived context" from whence it emerges and to which it relates. Signs, activating the imagination, cannot, unto themselves, guarantee communication. Though every passion naturally determines "the tone and movement of that music which is most proper for expressing it," its meaning must be obfuscated by linguistic habits. To adjust music to the "musical habits" of language, without encroaching upon its being an independent art, is what Condillac suggests as a remedy. Believing in the superiority of ancient music, he supported novelty and change in the art.

At any rate, music is distinct from language in the cognitive processes that it involves and in the kind of significations it conveys. Poetry should revert to music, rather than linger on ideation, for in the final analysis, poetry likewise tries to affect the "passions of the mind." On the whole, whereas Leibniz did not distinguish between the conceptual and the emotional, as far as his ideal language is concerned, Condillac did so. He sought "perfection" in each of these components separately, that is, exactitude in linguistic concepts, as they apply to the world outside, and exactitude in artistic means, as they relate to the world as experienced.[18] It is his implied division of the mind's faculties, despite his emphasis on the continuity of its operations, which brings Condillac close to Baumgarten's aesthetic propositions.

The Making of Culture

Renowned as the founder of Aesthetics, Alexander Gottlieb Baumgarten (1714-1762) started his career with the publication of some notes on Horace's *Ars poetica* (his *Reflections on Poetry* of 1735) and ended it with a perceptual theory in which art played a major part (his *Aesthetica* of 1750). From the very start he adopted the Leibnizian bifurcation of the mind into higher and lower cognitive faculties, assuming their analogical relation. Employing concepts central to Leibniz's and Wolff's epistemology of the scientific,[19] Baumgarten, in his aesthetic theory, also took into account the uniqueness of perception. Thus he acknowledged the *clarity* images share with scientific ideas, insisting nonetheless on their "fusion;" images do not permit internal distinctions, he argued, for they depend on a kind of continuity. Perceptual clarity is "extensive," that is, synthetic, pertaining to the ability to perceive, at one and the same time, as many components as possible of the sensate idea, whereas that of the conceptual is "intensive," that is, analytical.

It is the "lower faculty" which is the realm of poetry and all the other arts, argued Baumgarten, and not the higher one, as the French Classicists maintained, since it is the former which aims to reconstruct a sensate experience that activates the imagination. However, this faculty aspires to perfection no less than does scientific theory. It is this assumed perfection which justified the epistemological analogy he employed.[20] Naturally, norms and rules of poetic and artistic perfection are to be defined in accordance with their own confines. Aiming at construction rather than analysis, all that pertains to science and logic is reversed in art. Whereas science, with the aid of logic and a few concepts, reduces complexity to "simplicity," art makes the seemingly simple "flower." The foliage, however, evolves only through components *soundly* combined, otherwise they fail to capture the imagination.

The Leibnizian principle of "sufficient reason" is enlisted by Baumgarten in order to establish the regularity of the sensate. Unlike science, in sensate discourse it relates to the association of resemblances and the fulfillment of expectations. Since it is not easy to formulate cognitive principles, poetics should never be reduced to strict grammatical, logical or perceptual laws. The latter may, in fact, constrain the "license" which is at the heart of each artistic creation. Imitation in art, according to Baumgarten, is not that of nature, but of the effect it awakens. By opening the realms of art to the full array of the imagination, Baumgarten achieved a kind of parallelism between his world of fiction and the Leibnizian "possible worlds." All works of art, accordingly, turn, as it were, into full-fledged monads, that is, into organic units located in one of the possible worlds (i.e., artistic media) which define its nature. Though Baumgarten's essay on poetry appeared, outwardly, as still belonging to the *ut pictura* tradition, it went far beyond its premises without having spelt it out in so many words.

It is here that Baumgarten and Condillac meet again, since both ascribed to "sensate discourse" an immense importance. The mechanism of language, according to Condillac, efficient as it may be for the ordering of reality, quickly leads us astray from that which it intended to represent, for people become all too easily attached to words and composite figments, as though these were by themselves the substance of life (Condillac 1974: 120-2). The major fallacies of metaphysical systems result from distortions of language. In order to preserve truth-values, maintained Condillac, one has to return, again and again, to the initial level of language. Baumgarten, too, believed that sensate discourse leads to epistemological gain, though he still conceived it as related to metaphysical premises. In the investigation of the ways in which we conceptualize the world, the *ars pulchra cogitandi* participates in the attempt to reconstruct and comprehend the preestablished harmony which exists between man and the world, including all of its manifestations. The philosopher, maintained Baumgarten, like Shaftesbury (as we shall yet see), cannot overlook the investigation of art and the part played by the arts in the strengthening of thought and the enrichment of contents.[21] Though rationally oriented, both Baumgarten and Condillac, maintained Cassirer (1951), seem to advocate a kind of sensitivity, a sort of "tuning" of oneself and in relation to others.

Baumgarten did not treat music directly, but his treatise on poetry (1735) suggests a "correspondence" between music and poetry (Katz and Dahlhaus 1989 ii: 575). Dealing with sequence, Baumgarten tells us, poetry aspires to a unity to which its dynamic elements contribute. Time is, thus, introduced into the discussion, not only via rhythmical units but, as in music, via other dimensions of time. It was, however, the Horatian *lucido ordo,* rather than the Aristotelian imitation, which interested Baumgarten, who interpreted the former as a method in which the level of clarity increases as the piece progresses. Though concerned with the distinctness of poetry and painting, Baumgarten highlighted, indirectly, the new alliance of poetry and music.

It was left to Baumgarten's disciples, Krause and Sulzer among many others, to expound the relevance of his theory to music (see chapter 6). Whether directly or indirectly, many of Baumgarten's ideas reached the British isles and mingled with other ideas (Cassirer 1951; Engell 1981). By comparison, the impact of Condillac was more direct; his essay was translated into English as early as 1756. Though the translations of Dubos's and Batteaux's aesthetic treatises (of 1719 and 1746 respectively) preceded that of Condillac, the latter was undoubtedly of greater consequence to theories of art. Having gone beyond the classist principle of imitation, he seems to have provided theories of art a more fitting epistemological anchorage.

Indeed, with the influence exerted by Baumgarten, Condillac, and their followers it became gradually clear that the understanding of artistic realms depends on a theory of perception; at the same time it was increasingly understood that theories of culture are no less relevant. That only the intersection of two such theories may yield fruitful insights for the understanding of art dawned only gradually upon thinkers in the period. The man who most clearly understood this unique condition was undoubtedly the Neapolitan philosopher

Giambattista Vico (1668-1744). If Vico was not discussed until now, though his famous treatise *Scienza Nuova* already appeared in its two early editions in 1725 and 1730 respectively, it is because his work was not widely known at the time. Though Berlin (1981) tells us that he went "against the current," Vico's way of thinking points to an alternative route already made possible in the philosophical horizon of the time. As such, we believe, it has a bearing on much that was discussed hitherto.[22] As concerned as we are with the actual connections among our thinkers, we are no less concerned with their subsequent influence and, not less, in the entire paradigm which they helped usher.

Like Shaftesbury in his "doctrine of wisdom," which suggested a different kind of system of knowledge, Vico was impressed by the new historiographic and ethnographical explorations and tried to form a "new science." Like Leibniz and Shaftesbury he, too, was eager to comprehend the "wholeness" of mental activities, identifying their compartmentalization with "barbarism" (Verene 1981: 193-221). His central contention, however, that "truths are made," provided him with the base for his view of the sciences: Mathematics, he argued, is nearer to history than to physics, for in both mathematics and history man is the maker of that which he eventually recognizes as "true" (Verene 1981: 39; 54-6). Philosophy must be wedded to history, or more particularly, to philology, since it is in language that cultures are formed.[23]

Thus, Vico's *Scienza Nuova* contains data gleaned from philological discovery, fables and myths, customs and sacrifices, songs and economy. Its thrust, nevertheless, is philosophical. The uncovering of principles governing man's behavior as he molds and shapes his worlds, was Vico's main concern. It is primarily history, he believed, which may tell this tale. Hence, Vico set out to examine ancient primitive cultures and, to some extent, contemporary ones as well (Verene 1981 :70-73; 106-122). Historicizing man lent epistemology a new kind of perspective, separating the distinctly universal from accumulated knowledge, whether systematized or practical. Moreover, it liberated man as a product of fate or divine will and posited him as free creator of his own destiny, of his own humanity.

Creation starts from perception. In order not to become immersed in a "flux of sensation," whereby each new sensation conceals the previous one, the mind, argued Vico, fixes itself on a "locus"-- a point which is perceived as an image combining within itself the whole flux.[24] The complete process, however, is inconceivable without "memory." Vico, in fact, elaborated on memory, discerning within it three different aspects which together form an inseparable totality. Whereas *memoria* is simply the power "to bring to mind, to find in the here the not here and in the now the not now, *fantasia* reorders what has been recalled and shapes it in a way significant to the recaller. Through *ingegno*, formations are created, combining past acts so that they influence present ones" (Verene 1981; 103-5). We might say that Vico already populated perception with fictive cognitive entities—"psychological presences" the clinching of which is already guided by preconceived notions.

Vico insisted, however, that even at the perceptual stage the social consciousness is present, as it is with all that pertains to language, which is by definition a socially shared phenomenon. Elaborating upon the rhetorical tradition, Vico insightfully observed that thinking, as we know it, is primarily metaphorical (Mooney 1985: ch. 3). Metaphors, he argued, are categorically different from logical thinking and should not be regarded as "colorful or rhetorical embellishments of genera abstracted from empirical particulars" but rather as a "particular" concretizing of a "whole."[25] Thus he lent, in retrospect, legitimation to artistic practices of the seventeenth century.

Altogether, the mind, according to Vico, operates symbolically, transforming all sensations into a world of meaning, permanently seeking coherence and "truthfulness." It should be emphasized that it is not the human mind that changes; what changes are the cultural artifacts which are the creation of human consciousness. The enlightenment, accordingly, was misguided in its attempt to construct a science of human nature on the model of that of physical nature, for the only basis for a science of culture and a metaphysics of mind are historical investigations of the different encounters of human consciousness and nature, as they occurred at different times, situations and parts of the world. The boldest contribution in Vico's anthropological historicism, tells us Berlin, is the notion that there can be "a science of mind which is the history of its development, the realization that ideas evolve, that knowledge is not a static network of eternal, universal, clear truth, either Platonic or Cartesian, but social process, that this process is traceable through (indeed, is in a sense identical with) the evolution of symbols—words, gestures, pictures, and their altering patterns, functions, structures, and uses" (Berlin 1981: 113).

While Vico's type of historicism and social theory already exerted a strong influence in nineteenth century historical and social thought,[26] it is only in the twentieth century that it became possible to appreciate the full scope of his new science (Dorfles 1976). It is during this century that it has become clear that understanding science, in fact, any science, requires historical awareness which takes into account man's consciousness.

We have already witnessed, in chapter 3, how "truths were made" in the musical realm, that is, how musical signs acquired symbolic meaning. The theoretical notion, however, that the mind operates symbolically, transforming sensations into worlds of meaning seeking coherence, emerged only gradually in the deliberations on the arts. Attempts to better understand the relationship *among* the arts and the *uniqueness* of each seem, in retrospect, to have precipitated modern scientific notions of the arts more than did the inquiries into the relations between art and bona fide science.

Notes

1. This substitution theory will be called from here on "synonymous semantics." For this view and its fallacies see Eco 1984: 26-33.
2. Essences for Aristotle, are substances that gain in substantiality the more qualities they comprise. Aristotle clearly preferred to think of substance as essence. Whereas the distinction between class membership and class inclusion is not altogether clear in Aristotle's account, it gave rise to a number of interesting questions concerning the truly real in nature. Is the truly real known by inference or directly, objectively or subjectively, as part of the world of phenomena or of human understanding? All these questions were indeed asked during the period we are dealing with, and subjected to refined scrutiny, yielding interesting insights and enlightening arguments.
3. Of interest in this context are the theories and method of the fourteenth-century thinker, Ramon Lull, which the seventeenth century saw fit to resurrect. For Lull's theory and influence see Yates 1982. Lull, of course, is only one example of the impact occult traditions had on more "scientific" approaches to symbols and hermeneutics. See also Dieckman 1970.
4. As Summers demonstrated, clear, distinct, and confused ideas have important roots in classical and medieval thought, with which Descartes was partially acquainted (Summers 1987: 182-195).
5. Locke 1961, intro. 8; as quoted in Yolton 1985: 146. See also Adams 1971.
6. In this, as well as in his conception of lower and higher cognitive faculties, he followed Aristotelian concepts such as the "common sense" which relates to "the

sense of senses", along some of its important medieval (such as Boethius's and Aquinas's) and renaissance (such as Leonardo's) elaborations, as we learn from Summers (1987: 71-109; 182-197).

7. From a letter to Tschirnhaus, May 1678, as quoted in Ishiguro 1972: 35.
8. From *De organo sive arte cogitandi* we learn that for Leibniz the very ability of the mind to circumscribe a certain entity is dependent on the clear distinctness of the latter from that which is not it. (e.g., God and nothing, presence and absence) see Eco 1984: 23.
9. Modern thinkers identify Leibniz's binary condition—though without paying him tribute—as the infrastructure of syntax, semantics and computation, defining it either in terms of opposition or of negation.
10. For these ideas see his "Lingua generalis" from February 1678, published in Couturat 1961.
11. These dichotomous concepts are used by Langer in her informal talk entitled "The Art Symbol and the Symbol in Art" (Langer 1957). The Art Symbol, according to Langer, stands for the work as a whole, which is "the image of feeling... a single organic composition.." the elements of which are not "independent constituents as words are constituents of discourse." Symbols of the latter kind do occur in art, argues Langer, they have meanings which can be stated. However, their meanings are not part of the works' import, but "elements in the form that has import—the expressive form."
12. Interesting in this connection is Ross's (1984) understanding of LeBrun's insistence on iconic identity of expressive physiognomy, as a necessary pre-requisite for the *narrative* coherence of a certain picture, related to its overall structure (as exemplified in his interpretation of Poussin's "The Israelites Receiving the Manna in the Desert"), rather than to the natural and scientific status of certain facial expressions related to particular emotions. "The judicious contrast," writes LeBrun, "which gives movement and from which issues the different dispositions of the figures in which the situation, the aspect, and the movements conform to the story—creates the unity of action and the beautiful harmony which one sees in this picture" (LeBrun 1697/1883: 52, as quoted in Ross 1984: 42). Pointing to an aesthetic awareness often denied to French neo-classicism, Ross joins those, who like Borgerhoff (1950) pointed to its "freedom," and thus to its probable impact on British aestheticians.
13. Warburton's treatise was translated to French as early as 1744 (Dieckman 1970: 125) and later also to German. A similar approach is to be found in Bernard Mandeville's *Fable of the Bees*, but direct influence of both on Condillac cannot be substantiated.
14. Both quotations are from Hume's *Treatise* (1739), Book 1 Part I, Section 7 (Hume 1978, pp. 20 and 22 respectively).
15. Stam evaluates Condillac's emphasis on the "language of action" (gesture), as something of a novelty. He notes that the theme became important in aesthetic criticism, especially with Diderot (Stam 1974: 9).
16. For a similar modern view see Eco 1976, who describes such preverbal utterances as "undercoded messages" from which more codified ones are generated.
17. Condillac discusses the decomposition of linguistic gestures in the chapter "On Words" and onward (Condillac 1974: 237-264). His interesting ideas are echoed in modern thought. Contemporary discussions of the *signal* and the way it is understood, is a case in point (Neisser 1967). Attention understood as a process of *constructing* information is another (Gardner 1985: 120, 213). The relating of

means to pursuits of ends, as a process creating new mental entities, looms large since James (Gardner 1985: 108). Modern semiotics further explicates the nature of "gestural" signs (Eco 1976: 10; 118-20).

18. There is, however, an inconsistency in Condillac's argument. His inflexions of the voice are gestaltistic yet constrained by their constituent intervals. The choice of intervals, Condillac speculated, emerged from the harmonies associated with Rameau's *corps sonore*. The principle of pleasure is associated with these intervals, embodying, as they do, consonant relationship. Such a claim, however, reintroduces Descartes from the hind door, for it assumes a perfect mathematical base as a prerequisite for artistic perception. This fallacy, as we shall yet see, dominated Hutcheson's argument as well.
19. Though considered, together with Leibniz, among the founders of the German Enlightenment, Christian Wolff (1679-1754)—Leibniz's disciple and Baumgarten's teacher—adhered to the old scholastic framework more than Leibniz.
20. Cassirer (1951: 300) points to an important predecessor of Baumgarten precisely among the so-called French Classicist—Bouhours' treatise *The Art of Thinking in Works of the Intellect (La manière de bien penser dans les ouvrages de l'esprit)*, 1687. According to Bouhours, explains Cassirer "the value and charm of aesthetic appreciation do not lie in precision and distinctness but in the wealth of associations which such appreciation comprises." Moreover, the concept of *delicatesse* as used by Bouhours, maintains Cassirer, even amounts to a new organ. "The aim of this organ is not, as mathematical thinking, consolidation, stabilization, and fixation of concepts; on the contrary, it is expressed in lightness and flexibility of thought, in the ability to grasp the finest shades and the quickest transitions of meaning." The above mentioned study of Borgerhoff (1950) goes even further to show how this conception emerged from the insistence of the French on the *Je ne sais quoi* principle. Bouhours was especially attracted to the mystery of the emblems and Boileau and La Bruyère to that of the sublime. Franke (1975) discusses a more direct way leading from Leibniz and Wolff to Baumgarten.
21. The strengthening of thought, that is, the cognitive profit of the psychological activity to which both Baumgarten and Condillac relate, is something which Descartes has already emphasized in his *Passions of the Soul* (Descartes 1985).
22. Whereas the Leibnizian-Shaftesburian approach was subsumed in Baumgarten's aesthetic writings, one can find much of Vico's way of thinking in the writing of Condillac. However, whereas the historical connections between the former are clear and certain, the relationship between the latter are rather dubious. As both Wellek (1969) and Pons (1969) argued, the ideational proximity which prevails between Vico and Condillac, probably issued from their shared problematics and shared sources rather than from an actual reliance on each other. Yet Warburton could have served as a link between the two (See Dieckman 1970: 128). In contradistinction to the above scholars, Costa (1976: 247-9) believes that Vico did exert some influence in his own time.
23. Mooney (1985) also emphasizes this point, connecting it to a long tradition, which investigated the human mind through language.
24. It is here, according to Mooney, that rhetoric and cognition converge in Vico's thought, for the former (as a conception, not as a historical discipline) supplies the *topos* which the latter subsumes (Mooney 1985: 82-3).
25. See Mooney 1985: 75. It is no wonder, given Vico's entire philosophical outlook, that his view of metaphors is similar to a modern one, exemplified, for example, in Black's (1962) theory discussed in chapter 2 above.

26. Vico's type of knowing, which is founded on memory and imagination is a knowledge resting on sympathetic insights. As such it influenced the trend in historical writing which emphasized a kind of identification (*Einfühlung*) with the historical moment investigated (Berlin 1981: 117). His social theory, which recognizes stages in the development of society related to different patterns of social organization, seems to have influenced both Hegel and Marx.

5

Sense, Sensibility and Commonsense—The British Paradigm

The British School

In their joint efforts to achieve an "unsuspected eloquence," text and music were destined to adjust to each other's caprices.[1] Though harmony was not always the immediate result, the process led both arts to acquire elements and procedures that were not initially inherent in each.[2] Thus, in the celebrated remarriage of words and music in early opera, it was the "musicalized" pastoral that served as libretto, whereas music was enlisted to highlight the rhetorical qualities of the text.[3] This symbiotic trend gained momentum in the course of the seventeenth century until the well-known *serva padrona* debate of the ensuing century, concerning "who served whom" in the alliance between text and music in the opera, reopened questions about the distinctness of each of the arts. It was eventually left in the hands of those who realized that the whole issue required a serious review. As it turned out, it was their deliberations which precipitated the shift from the well entrenched *ut pictura* to the *ut musica* paradigm.

The British contribution to this momentous change is too often overlooked, and the shift itself is underestimated. Draper (1932-3) pioneered in raising the issue anew in modern research, attributing the shift to an antecedent phase of Romanticism. Draper also brought the British group to the awareness of scholars. Later scholars disagreed on various points, including different evaluations of the uniqueness and importance of this particular phase. Thus Rogerson (1945) spoke of an "English reaction" to French neoclassicism, whereas Schueller (1947; 1948; 1953) considered most of the English discussions concerning music "anachronistic" from a philosophical point of view. Equally inclined to underestimate the English contribution, Abrams (1953) emphasized instead the contribution of the Germans, thereby missing out on a major source of support for his own theory. Kristeller in his "Modern System of the Arts" (1965) repeated the same view, acknowledging, however, that the English writers in the second half of the century delved into concepts and principles concerning the arts more philosophically.[4]

In recent years, the analytical philosopher Peter Kivy (1980, 1984, 1988) found new interest in some of the British writers; not treating them, however, as a unique *group*, he, too, overlooked some of their innovations. Contrary to these views, we insist that there is no room for doubt over the importance of the shift introduced by this group of writers, nor over the significance of the British attempts to fathom the "correspondences" among the arts as well as the uniqueness of each of them. These attempts were connected with intellectual queries of a much broader nature, the importance of which has not diminished to this very day.

True, some of the British treatises, though not the representative ones, are somewhat "old fashioned," as has been alleged, yet the contribution of the group as a whole, we insist, may rightly be regarded as a major theoretical breakthrough. The broader the historical perspective, the more prominent the British become. A broadened view reveals, for example, that the British treatises of the eighteenth century show explicit awareness of the validity of the pious intent of the Florentines (towards the end of the sixteenth century) to unveil the communicative powers of music, and that their daring hypothesis could be anchored philosophically. What needed unveiling, therefore, were the particular *ways* by which this was achieved. A broader view also reveals that the British inquiries into music and the arts concerned the "*operations of the mind*" which artistic realizations involve, more than they addressed the artistic objects themselves. In their desire to decipher the function of symbolic systems, in their attempts to understand what had nowadays been termed "world *making*,"[5] the British theorists were among the initiators of a new approach. Guided by the then newly established separation of the fine arts from the sciences, they tried to discover deeper significance *within* the arts. Their type of inquiry eventually embraced science as well, leading to new alignments between science and the arts.

Though the "modern system of the arts" constituted an integral part of the thinking of our British writers, their treatises do not yet contain the philosophical systematization characteristic of the great aesthetic theories of the following century.[6] Even if Kristeller is right in assuming that the very establishment of the new system meant recognition of the epistemological uniqueness of the arts, eighteenth-century attempts to conceptualize that uniqueness were still sporadic and largely unsystematic. The "new system," at any rate, is better understood as a historical culmination, entailing conclusions drawn from a long process of development *in* and *among* the arts.[7]

Process more than event applies equally to the *ut musica poesis* discussions, to the point of eclipsing the novelties contained in the newly undertaken "comparisons among the arts" of the British group. Without forgoing considerations pertinent to the specific moment, the breakthrough of the British treatises, however, can be fully appreciated only through a retrospective analysis of the relevant theoretical queries that preceded them. Accounting for the very appearance of the numerous new "comparisons," let alone the issues they raised, requires such a wider base.

Our retrospective analysis revealed that the theoretical background which lead to the British treatises followed, in the main, two different kinds of deliberation. The one was closely linked with the search for what the arts have in common (a search that led to the establishment of the "modern system of the arts"), which drew its standards primarily from the humanistic tradition, attentive, nonetheless, to the requirements of an ever increasing amateur audience. Bearing the stamp of a free, unfettered discussion, it was basically commonsensical. The other kind of deliberation was closely linked with attempts to clarify the epistemological status of symbolic systems discussed in the previous chapter. It was an attempt to achieve greater precision through an alliance with systematic philosophical thought. Within the first kind of discussion, it was, of course, the comparisons among the arts that interested us most. In fact, the British predilection for "common sense" made the bridging between the two approaches possible. Of course, in order to explain why the British, of all the European peoples, were so inclined and prepared to deal with the subject, one must take into consideration the cultural conditions and the atmosphere which characterized the British scene.

In what follows we shall try to introduce the group whose contribution was decisive for the changed conception of art in the eighteenth century, particularly that of music. It is this new conception, we believe, that paved the road towards modern art, a road which is inconceivable without the hypotheses bequeathed, by our group, to the field of cognition in general. Accordingly, this chapter aims: (1) to reveal the underlying connectedness among the ideas expounded by the British writers we have grouped together, and (2) to trace the growth of thought that led to the separation of the aesthetic from the knowledge of the métier and, eventually, to the undermining of semblance as prerequisite for verisimilitude.

But before we undertake this twofold task, a short general portrayal of the group is in place. It is evident that all of our heroes enjoyed a classical education. Though they treated the authority of the ancients critically, they all accepted, as it were, "the great chain of being," which highlighted the cumulative aspects of culture and its internal hierarchical structure. Likewise, they all embraced those cultural achievements that had risen, in the course of time, to canonic status. All of this was upheld by a worldview that was still largely based on postulates in which the particular was inferred from the general. Although this worldview, as we have seen, had already undergone significant change, by their time, the transformation in thought had not yet infiltrated the public domain.

The growing attention to cultures, other than the European, however, may have had a greater share in changing modes of thought than did the slow diffusion of philosophical deliberations, for much that had been taken for granted as universal became with the increased ethnic awareness, *demonstrably* relative. Clearly, the need was strongly felt for a better understanding of man-made culture and, for that matter, for better formulations of the naturally given.[8] The relationship between the two—the cultural and the natural—was also expected, as a consequence, to undergo change. Conceptions do differ, it was agreed, but what about the very acts of cognition—conceiving and perceiving? In what way are the two related to each other, and wherein do they differ? Questions of this kind became salient, not least because of the widened horizon of the "real" world, which could no longer be contained within the frame constructed on the basis of the earlier postulates. Indeed, the old frame was forced to yield to laws inferred from careful investigations of particular instances. A more flexible frame arose, one that could be expanded to hold new revelations, that is, that would contain the *potential* for extension into further areas of thought and meaning.

Changes of this magnitude do not occur overnight. Our heroes, as stated, were carriers of a venerable tradition, vital to this very day. If their clarifications were of momentous significance to the understanding of art, it is because they were steeped in culture, committed to its understanding, and attentive to all that could enhance such an understanding. Interestingly, the very people who challenged the *ut pictura* legacy made ample use of the pictorial tradition, from Leonardo to Caravaggio, which was concerned with realistic illusions. This, in fact, enabled them all to come to grips with the intricacies entailed by the process of imitation.

Though they held much in common, the members of our group differed in important ways. Taken together, these differences, no less than the similarities, tell the story which we have taken upon ourselves to tell. Over a span of several decades, they formed a kind of "invisible college," dedicated to the clarification of key issues concerning the arts, issues which had become salient in the world to which they belonged. Like the invisible colleges of science, they, too, created over the years a community of discourse, based on an acquaintance with each other's writings. They were familiar with each other's ideas either directly, having read the treatises, or indirectly, through mediating members whose treatises they had read. Naturally, those who came last could benefit from all those who preceded them.

In order to accomplish the double task of this chapter, we present the group in chronological order, introducing an internal classification among its members. This procedure is expected to highlight the cumulative aspect of their deliberations, while pinpointing crucial stations in its development. We divided our group of writers into four "generations," in historical order. Shaftesbury and Hutcheson are included as members, because they seem, in retrospect, to have provided cornerstones for an edifice the completion of which continued to challenge the following generations.

In the generations following Shaftesbury and Hutcheson not everybody was a "professional" philosopher, but the nature of the discussion involved a great deal of philosophizing. The fact that non-philosophers also participated in the discussion was clearly to its benefit. Interestingly, each generation included a bona fide philosopher, a musician and a man of letters. At times these roles overlapped and were embodied in a single person; at other times they were distinct. In the second generation it was Oliver North, the musician, who provided the knowledge of the metier, James Harris, the Aristotelian, who provided the philosophical framework and Hildebrand Jacob, a poet and translator, who delineated the cultural horizon. These three did not communicate directly with each other; were they to do so, they would have talked at cross-purposes. While in the second generation the separation between the musician, the philosopher and the man of letters was distinct, these distinctions became blurred in the third generation, not least because of the understanding that was reached in the second generation. It is of note that two of the members of the third generation—Beattie and Webb—can be considered men of letters as well as philosophers, and Avison, the musician, can also be considered a man of letters. The last generation, however, reveals that profound philosophical discussions of art must also involve an intimate acquaintance with the arts, themselves, and their cultural environment. Thus, in closing the "circle," it is not surprising to discover that Twining should have drawn his musical knowledge directly from *the* musical authority of the day, Burney, himself a direct follower of Avison and North, while Smith, the moral philosopher-economist, should be linked to the philosophical school of our "founding fathers." Thus, we start with a clear philosophical orientation and seem to end with one.

The importance of shared legacies and worlds of knowledge notwithstanding, the power of the group also stems from the specific dowry brought by each of its members to enrich the "family capital." The knowledge each brought separately can be seen to have influenced their own investigations and unique conclusions. Once circulated, however, these turned into arguments with standings of their own and established starting points for subsequent writers, enlarging the entire scope of the deliberation. Jacob, for example, contributed two important "fortunes"—contemporary French aesthetic theory, with which he seems to have been well acquainted, and musical theory, from Guido to Zarlino, with which he was likewise familiar. The first was decisive for his challenge of the *ut pictura* paradigm. He was among the first to deal with aesthetic and artistic criteria in a nondoctrinaire way, taking into account the constraints of the medium, on the one hand, and that of history and culture, on the other. The history of art, he tried to convey, is a product of successive choices, circumscribed by previous choices. His understanding of the interaction between theory and practice in the history of music, although lacking in substance and accuracy, was central to his entire position.

In like fashion, one can show that Beattie's acquaintance with the ongoing literary genres of the time, especially the newly created novel, deepened his understanding of the aesthetic. The realistic materials of the novel brought to the fore certain qualities which separated artistic vision and artistic choices from the ordinary perception of reality. Webb, by contrast, instilled life into treatises that were "buried" from the outset, because of their peculiar mix of the occult and the scientific, so typical of seventeenth-century encyclopedists. Considering that he himself tried to bridge incommensurabilities, this does not come as a surprise; moreover, it is these treatises—Kircher's, Vossius's, and others—that provided the tools which he put to good use. What follows, then, is an exposition of the continuity of contemplating the arts, especially music, over four generations. We begin with the foundations laid by Shaftesbury and Hutcheson, continue with North, Jacob, and Harris. Avison, Beattie, and Webb, belong to the third generation, leading finally to Twining and Smith.

Neo-Platonism Emanates, Empiricism Transcends: Shaftesbury and Hutcheson

Morality and Beauty Governed by Perfection

Shaftesbury was essentially a neo-Platonist, like many others who delved into aesthetic queries from the middle of the fifteenth century onward. Though Aristotle left his mark on aesthetic theory, especially after the translation of his *Poetics* in 1498, Platonism and neo-Platonism still dominated aesthetic discourse, especially in the seventeenth century. Over the course of the eighteenth century, however, Aristotelian thought made increasing inroads into aesthetic notions that were initially Platonic. Why Platonism, to begin with, was more conducive to theories of art—given Plato's negative attitude to the arts in his Utopian State—is a question that goes beyond our immediate concerns. Still, entertaining Aristotelian thought in the attempt to bestow upon theories of art greater philosophical rigor, raises the question anew with regard to Plato: Why is it that most of our writers, who may be characterized by a decisive Aristotelian bent, looked to Shaftesbury as the initiator of a new approach in the contemplation of the arts?

By the time "perception" gained the attention (for which Locke, as we saw, was largely responsible), the subjective could no longer automatically be attached to the objective. The subjective, thus, called for further qualifications; neo-Platonism, which Shaftesbury still adhered to, did not even differentiate between the qualities of the object and the ability of the subject to perceive them, taking for granted the preestablished connection between the two. A close examination of Shaftesbury's neo-Platonism, however, reveals some crevices in this conception. While outwardly rejecting Locke's legacy, he could not help but be challenged by it. Locke, after all, was Shaftesbury's tutor in his formative years (see *AM*, introduction to Shaftesbury).

Shaftesbury followed Plato in claiming that the mind represents a culminating point leading from the inanimate to the animate, from matter to spirit. However, within this hierarchy of increased rationality, Shaftesbury introduced an element of a-rationality, which is not opposed to the rational but interacts with it in particular ways. In this, he was a child of the seventeenth century, which introduced a differentiation between rational and affective behavior as well as between primary and secondary qualities. While Hobbes emphasized the negative aspects of the affective behavior of man and Locke was primarily occupied with secondary qualities, Shaftesbury was ready to see in the affective, unlike Hobbes, the source of the benevolence of man, attaching to it the domain of secondary qualities through the idea of beauty.

Shaftesbury understood that only by psychologizing benevolence, the field of ethics and ethical behavior may be rescued from Hobbes' negative views. This lies at the base of the Shaftesburian Moral Sense which is governed by the notion of perfection, itself not possessed by specific contents. Beauty, likewise, governed by notions of perfection, is conceived by Shaftesbury as an independent sense. In comparison to morality, beauty is more readily experienced by ordinary people, and exemplifies what Shaftesbury, the moralist, had in mind in the field of ethics. Shaftesbury thus turned the neo-Platonic "divine illumination" into human senses (Townsend 1982: 209)—what Hutcheson later called the sense of beauty and the moral sense. It is in his attempt to justify his position with regard to the standing of the "senses" that Shaftesbury diverged from pure neo-Platonism, raising major philosophical problems.

In its notion of perfection, Platonism assumed not only a mind capable of grasping perfection, but also qualities related to the object itself. This is, in the first place, why neo-Platonism was so conducive to theories of art. Unlike former neo-Platonists, however, Shaftesbury was not content to leave the notion of perfection (in art) at the ideational level, but wished to unveil its realization in *real* works of art as well. By disclosing their underlying rules, Shaftesbury believed that works of art could provide models for future artistic creation. Originality and change are not encumbered by such models; on the contrary, he believed they would benefit directly from the sound *principles* which these models exemplify. Shaftesbury's "perfection," thus, integrated the cultural and the "preestablished," relating both to the values of a moral society—his main concern. Indeed, with Shaftesbury neo-Platonic Perfection emanates, while sensual empiricism transcends. Judgment links the two. Judgment, however, is at all times dependent on *normative* points of reference, which are themselves culturally determined. In the final analysis, Shaftesbury's involuntary sense of beauty already contains some reflective aspects. Like Vico and Condillac later, and unlike Locke, Shaftesbury sensitized us to the fact that within the subjective itself it is not easy to separate the cultural from the sensual, the learned from the given.

Shaftesbury's concern with the public good made its impact, above all, on a middle class that rose to status in matters of culture. His ideas may be interpreted, as Caygill tried to do, as a benevolent formulation of a commercial credo, according to which "merchants are exchanging according to individual interests, and finding these interests disposed toward the end of the public interest" (1989:51). The economic equilibrium of a free market, attributed by Shaftesbury to providence, will be transformed, by Smith, into a guiding, though invisible, hand. So will the notion of benevolence become secularized in Smith's aesthetic theory. Nonetheless, one should bear in mind that in the almost seventy years that separate their main writings, Shaftesbury's integrative theory, no matter what its "ulterior motives" might have been, introduced a new notion of the commitment of society to culture and of the responsibility of culture to the melioration of society. At the same time, he was well aware that only a society that rises above basic needs can indulge in the "luxuries" that concerned him. The concept of disinterestedness, so strongly identified with Shaftesbury, should be understood in light of the above, that is, not as a lack of interest, but as a *shift* in interest, from utility in favor of spiritual gain.[9] Once introduced, this shift created a separation between the moral sense (which was transfigured by Kant into *"practical reason"*) and that of beauty, a separation in what initially aimed to be a *unified* theory of values. That this new split had far-reaching consequences, especially for the secularization of culture and thought, goes without saying. By conceptualizing the prerogatives of the few—the educated and privileged—as a service to the many, Shaftesbury sought to promote culture as a commodity necessary for an improved quality of life. Enlightened societies tended to agree.

Important as Shaftesbury's contributions were, they could not save him from philosophical attacks. Indeed, Shaftesbury confused sense and idea, *telos* and the nature of being, and got entangled in circular arguments. As Caygill (1989) rightly points out, Shaftesbury posited the ultimate reality of beauty and virtue and claimed that individuals embody them. At the same time, he expected individuals to live up to them, knowing full well that they can act otherwise. Mandeville's famous *Fable of the Bees* (1723) attacked Shaftesbury on this very point. It is for reasons of this sort that Kant considered the moral sense to be, altogether, an unphilosophical conception. Shaftesbury, however, could not be bothered, for he was primarily concerned with the realization of the *potential* embodied in man. While Vico thought that man was able to understand only that which he himself created, Shaftesbury dealt philosophically with creativity. If criticism loomed large in Shaftesbury's arguments, it is because he believed that the development of culture should be subjected to checks and balances, which are themselves attained through a process which is itself creative. Like artifacts, criticism is a kind of "bringing about," in the Aristotelian sense, and not a mere reflection of that which exists. The interaction of artist and critic is essential to the relationship of action and reflection, creating the proper community of discourse that guarantees that the public is well served. That the artist should be viewed as a creator was already recognized in the course of the Renaissance, but that *reflection* upon art should also be viewed as a creative act, was a novel idea, which emphasized the "world-making" it itself involves. Henceforth, the fathoming of creativity will occupy a new place on the agenda of aesthetic inquiry. The special mix of neo-Platonic creativity with Aristotelian mimesis, so beautifully exemplified in the above arguments, will receive further clarification in subsequent attempts to unveil the cognitive processes it entails.

Still, though Shaftesbury broadened the field of aesthetic education to include culture as supplying the criteria of judgment, he basically left the *Je ne sais quoi*—the "I know not what" makes this artwork/ landscape beautiful/sublime in my eyes—unanswered. In fact, he failed to grant proper recognition to the answer that he himself provided. As Leibniz recognized at the time, "taste" cannot be dealt with in philosophical terms alone. (see *AM*: our fn. 10 to Shaftesbury's treatises). That history, sociology, psychology have a bearing on such matters, was, as we saw, profoundly explicated by Vico. Shaftesbury's insights still lacked the recognition of their far-reaching implications; he was unaware of the fact that philosophy's inability to unravel them constitutes in and of itself a philosophical position.

Computing Unity amidst Variety

Hutcheson is generally considered Shaftesbury's immediate follower; he himself acknowledged his indebtedness. While tightening some of Shaftesbury's arguments, he also made intensive use of the theory introduced by Locke. Though it may appear as if the hands were those of Shaftesbury and the voice—that of Locke, he employed the latter more tactically than strategically. Hutcheson, after all, was primarily a moral philosopher, concerned with the regulating forces that apply to society, affecting the cohesion of its individuals. In the final analysis, as he tells us, he was interested in the happiness of Man; the clarifications of truth provided a means toward that desired end.

Hutcheson's main attempt was to rid the moral sense, and the sense of beauty, from contents which Shaftesbury attributed to them, in his attempt to bridge the "preestablished" and the culturally acquired (see *AM*, our introduction to Shaftesbury). In so doing, he tried to defend Shaftesbury from the attacks launched against him concerning the *innateness* of the moral sense. Following Locke's differentiation between sense and idea, he enlarges the former to incorporate, what he called, the internal senses, underscoring thereby their reflexive character. "Immediacy," thus, surfaced as a major property of the aesthetic. However, unlike the immediacy attributed to the "external senses," the immediacy of the "internal senses," according to Hutcheson, is characterized by a sort of computation, which bestows unity on variety. The old precept of "unity amidst variety," of which the scholastics had made use, was now incorporated into the processing of the mind, reminiscent of Leibniz's "unconscious calculation of the soul." Like Leibniz, but even more directly, Hutcheson attached to the reflexive calculation the notion of pleasure, which supposedly safeguards the taking place of the computative action.

The pouring of Platonic contents into Lockeian vessels raised problems which Hutcheson had to overcome. By separating the reflexive from the reflective—the immediate from the contemplative—Hutcheson sought to explain how that which one reflects upon emerges as a prior independent mental image. Hutcheson believed he had solved the problem by relegating the latter to a special sense. Moreover, by defining the inner sense as a "power determined to be pleased," Hutcheson altogether dissolved "the difference between power and form, since power which is determined is already formed" (Caygill 1989:54). Hutcheson, thereby, found himself again in the domain of ideas, the intervening factor of pleasure notwithstanding. As with Shaftesbury, Kant was aware of the circularity of Hutcheson's argument, arguing that it "introduced a new sense as a ground of explanation," loading it with objective properties which it could hardly sustain (Kant, *Reflection* 6634; see Caygill 1989:53).

Pure rationality, however, was not what Hutcheson was after, for he wished to take into account also that which *defies* logical analyses. More than Shaftesbury, he was willing to limit the universality of aesthetic principles to necessary conditions only. In other words, why a pagoda should differ from a campanile, though both, according to Hutcheson, embody unity amidst variety, is not essentially a philosophical question. One might add, that neither is it merely a social or cultural one, for the separation of sense and idea, form and norm, is not a simple matter. Hutcheson, however, did not attempt to unravel these matters, though he sensitized us to them. Cognitive studies, even nowadays, aware of the difficulties involved in delineating the border between perception and cognition, all too often derive hypotheses with regard to the former via behavior related to the latter.

Given the enormity of the problem, Hutcheson, nonetheless, contributed something towards its clarification. Following Locke, he differentiated between simple and complex ideas, and it is with regard to the latter that he conceived the phenomenon of beauty. As Emily Michael (1984) points out, beauty, according to Hutcheson, is a "property of properties"; as such, Hutcheson revealed awareness that the experience of beauty involves different levels of perception. Had he followed this line of thought, he might have discovered that what he considered as a rudimentary level of perception, is already complex.[10] Furthermore, by conceiving "immediacy" as relating to the parts that constitute the whole, that is, unity, Hutcheson did not entertain the possibility that the parts themselves may be sorts of unities. "Immediacy" mistakenly functions, here, as the *cause* of unity, whereas in fact it is a consequence thereof. Still, as a property of properties, Hutcheson could more readily relegate beauty to the inwardly given, rather than to the attributes of the external object. Were we to concentrate, theoretically, on the object alone, we would fail to account, according to Hutcheson, for the fact that nonhuman creatures are not necessarily endowed with the cognitive properties that enable man to perceive beauty.

Hutcheson's emphasis on the computativeness related to properties, highlighted the *processing* of the mind, yet his understanding of "unity amidst variety" solely in arithmetic terms did not go far enough to explain the processing of aesthetic matters. In present-day terminology, one may say that Hutcheson tried to bridge image processing and propositional processing, but only the propositional factors were derived, in the case of Hutcheson, from external properties. Moreover, "unity amidst variety" as conceived by Hutcheson, does not go far enough even to explain the external, which harbors additional factors—quality as well as ideals. Hutcheson, as it were, treated the gestalt as a summation of its parts, overlooking its own uniqueness; it is the latter, after all, which he attempted to define.

That "unity amidst variety" does function, on a meta-level, for Hutcheson, becomes clear from the fact that it includes the concept of "relative beauty." Relative beauty, for Hutcheson, covers the whole field of resemblances, comprising "mimesis," rhetoric and a quasi-theory of intentions. Hutcheson, unwittingly, subjugates Aristotle to Plato, granting Aristotle aesthetic criteria, which Aristotle himself did not supply. In the final analysis, what concerned Hutcheson was the notion of coherence, similar to that which preoccupied Baumgarten in relationship to all that aspires to artistic standing in the lower cognitive faculty. But whereas Baumgarten supplied criteria by which to assess coherence in the artistic object *itself*, Hutcheson, like Dubos (from whom he might have learned), argues for a dependence of the artistic work on the object it imitates. Relative beauty, argued Hutcheson, "is founded on a conformity, or a kind of unity between the original and the copy." Nonetheless, Hutcheson did understand that resemblances constitute an essential factor, which has to be reckoned with in all that pertains to art. Although he left its philosophical standing in a state of ambiguity, he rescued "resemblances" from the abuses they suffered in the seventeenth century. It is interesting to note that in modern theories of perception resemblances loom large. Their standing in philosophical theories, however, is still dubious.

All of the above also explains the "what" and the "why" in Hutcheson's ideas concerning music. Hutcheson reconciled Zarlino and Galilei, harmony and affective import as artistic ideals. He was unperturbed by competing artistic desiderata, convinced that they can exist side by side, under that wide roof which is able to unify all kinds of varieties. However, what he understood by harmony does not go beyond Zarlino and Kepler, despite the advent of Rameau. With regard to the competing theory of the inflexions of the voice, handed down by the "school" of Galilei, he *did* introduce some novelties, trying to depart from mere metaphorical talk. Thus, he transformed the old concepts of reverberation (between sympathetic strings) and agitation (of the responsive heart) into the overall concept of resemblance, resulting from the ability of the mind to discern a well-established association (between a certain melodic line and the inflexions of the voice of an agitated person). However, he left the task of specifying the *conditions* that underlie this well-established association to other thinkers.

With regard to the concept of sympathy, Hutcheson thought he had solved the problem by fusing its physicalistic and figurative aspects (between which it hovered in the seventeenth century) into a cognitive concept, stating that "when our ear discerns any resemblance between the air of a tune to the sound of the human voice... we shall be touched by it... by a sort of *sympathy* or *contagion*."[11] Given his unclarity with regard to the aesthetic standing of resemblances, it is hardly surprising that he should have emphasized the sympathetic experience at the expense of an explication of its aesthetic conditions. So long as mimesis was treated as an overall theory, placing emphasis on the imitated object, one tended to dwell upon the affective response, rather than on the ways and the hows by which imitation is achieved from an aesthetic point of view. Hutcheson, nonetheless, introduced music into a contemporary discourse that dealt, in quasi-philosophical terms, with similar matters in the domain of literature.

Between the Particular and the Universal

The impact of Shaftesbury and Hutcheson on our British group can be reformulated in terms of the notions that emerged in connection with the "inner senses," namely, the *relations* between: the reflexive and the reflective, the moral and the aesthetic, the private and the public interest, order and perfection, the particular and the universal. These constituted new criteria with which to tackle the procedures of judgment in general, including artistic judgment. Judgment occupied center stage in the aftermath of the scientific and the social revolution, which England had undergone.

To be sure, in the history of civilization, judgment has always loomed large, but never before was it so intimately connected with the idea of beauty. Though Aquinas had early isolated beauty as a guide for proper judgment, it remained part of a transcendental "measuring" cause. The increased artistic production in the Renaissance, shifted the responsibility of measurement to *man*. God, according to Pico, delegated to human freedom this responsibility, causing, as

Caygill points out, "a complex relation of legislation and production" which succeeded "the old transcendental philosophy" (Caygill 1989:16). By severing measurement from responsibility, Descartes even undermined this relationship. These high points in the historical development are not isolated phenomena, for they encompass the entire transition from the "strain of living in a mysterious universe," subjected to the notion of revelation, to the emphasis on nature, itself embodying order, unity and proportion, that needed and could be unveiled.[12] This transition, moreover, was accompanied by a transition from contemplating God to the god residing in man.

This whole transformation constituted, in fact, the agenda of the Cambridge neo-Platonists (Patrides 1970; Powicke 1971). Design, order, and law became inseparable for them. To please by *rules* eventually became akin to what pleases, and what pleases must have an underlying rule that does not contradict nature, but, rather, reflects its laws. The "back to nature" ideology, initiated by Rousseau and fortified in the romantic era, makes it difficult to understand a historical moment in which to be civilized amounted to being natural. Ostensibly, the problem of judgment was resolved thereby, since the criteria employed for judgment were the very criteria that were already embodied in undisputed cultural values. Such a process, however, could take place only in a society that was complacent about the state of affairs it had already attained. The civil society achieved in post-Restoration England, succeeded to limit executive power and regulate both public and private finances. Eager to protect its own achievements, it had to balance social and economic forces, on the one hand, while granting them freedom, on the other. Needless to say, such a tall order distorted some issues and left many unresolved.[13]

If Shaftesbury still believed that the pleasure of beauty and virtue, which are in themselves sorts of *passive* contemplation, is suffice to navigate the welfare of the society, it became absolutely clear for Hutcheson, in his later writings, that the sheer contemplation of proportions won't deliver the material infrastructure needed for the very existence of society. This was precisely the basis of Mandeville's argument, which emphasized the *active* desire for material objects underlying not only industry but even societal order. Smith's "invisible hand" is based on an "exchange" theory that governs the *vari-*

ous strata of society, from which, according to Smith, everybody stands to benefit. But according to Hutcheson, production had to be enforced, because disinterested pleasure cannot overcome the selfish passions of those who have to deliver the basic goods. Being too shortsighted, the productive class cannot envisage its "obligation" to society at large.

Needless to say, this kind of thinking had far-reaching social implications. Even if slavery is something that Hutcheson, by "logical derivation," was able to entertain, he did not succeed to unite, theoretically, that which triggers the incentive and that which underlies the contemplative, not even in the domain of art. It seems that here, too, Hutcheson and Shaftesbury shared a problem that they were unable to resolve. Social forces, apparently, like aesthetic dictates, have to be contemplated first within their own proper historical and political frames. That does not mean that they do not affect aesthetic dictates; rather, the contrary is true. But in order to understand their interplay it is necessary to isolate first their constituent elements. In whatever form, constituent elements must be made salient.

On the whole, the issues raised by Shaftesbury and intensified by Hutcheson, did not yield definitive solutions to the problem of whether sense and idea were separate from each other, or whether in their interconnected form they could be subsumed by a unifying theory. Nonetheless, they directed attention to the problem of coherence vs. meaning, both central to symbolic theories, though coherence is supposed to have universal standing, whereas meaning is culturally circumscribed.

Media, Culture, and Mental Exercises: North, Jacob, and Harris

The separation of the aesthetic from the cultural, so necessary for the clarification of the arguments discussed above, is not easily achieved; it requires, paradoxically, a high awareness of matters of culture. Unexpectedly, such awareness is often found among those who had the opportunity to gain insight into the essence of an art while practicing it. However, the conceptualization of essences is necessarily dependent on theoretical frameworks of a more abstract nature.

The Making of Secular Musical Culture

The writings of Roger North (1651-1734) on music beautifully exemplify this point. A lawyer and architect of distinction, North had a fair education in mechanics as well as in philosophy. As a conscientious practical musician, he was able to extrapolate from his vast experience with music some speculative thoughts with far-reaching implications. A follower of the tradition of Morely, rather than that of a Fludd or a Vossius, he detached himself from the metaphysical, making room for more down-to-earth theories. Though he did not altogether cast aside the neo-Platonic theory according to which all the arts seemingly deal with the same subject matter, he revealed an awareness of the uniqueness of each of the artistic media. Consequently, it does not come as a surprise that Aristotelian ideas should have invaded his thinking.

North exemplifies the *climate* of the period, more than he does the rigor of its aesthetic formulations; for this reason, his treatise is not included among the treatises we dealt with extensively. Cultural climates, however, are no mere phantoms; they emerge from the discourse of people who are able to communicate with each other on the basis of shared notions and assumptions. These shared features lurk everywhere in the insightful pages of Roger North; even if they are never treated in a satisfactory fashion, they enable us to gain insight into the concerns that were "in the air." That North should rightly be considered a representative of his period is reinforced by the ways his writings were subsequently viewed. Burney, for example, in his historical writings on music, refers to North as a representative of his time, and an important author of musical investigations. Nor is it altogether impossible that some of our writers should have been directly familiar with his not-yet-published writings, which enjoyed a fair amount of circulation. This may hold true especially for Jacob who provided us with a "middle-range theory," one which stayed closely attuned to the surrounding culture. At any rate, North was in close contact with many of the important musical figures of his time and an avid participant in its musical life. The respect he commanded was in no small measure due to his thorough knowledge of the musical métier, as he himself tells us.[14]

The writings of North reveal, above all, the extent to which certain ideas were taken for granted at the time: The so called "modern system of the arts" was taken for granted; poetry, painting, and music were "naturally" allied, as was the notion of imitation as their binding factor. North, however, was among the first, in modern times, to apply the Aristotelian mimetic dictum to instrumental music, in which the very concept of imitation was challenged. By applying imitation to a sonata of Corelli, whom he admired, he combined the Baroque understanding of the singularity of affect with the Aristotelian theory of "action" (translated into movement), albeit in a simplistic way. Although primarily a "grammarian of music," North had an inquisitive mind, ready to speculate beyond practical matters in an attempt to embed the "science of music" in a larger theoretical framework.[15]

In his various theoretical endeavors, North tried to account for musical qualities—not local-contextual qualities, but all-encompassing ones—in an attempt to explain the *coherence* of musical works. He was ready to forgo much appreciated musical detail—ornamentations and the like—for the sake of "artistic unity." Indeed, the Aristotelian mimetic principle served North in a triple sense: as a guide for composition, as a base for apprehension, and as a point of departure for aesthetic judgment. It is musical perception, however, which constituted the major theoretical issue he grappled with. Among the many questions he posed and tried to answer, one finds the following: Given that music depends on transformations no less than it does on repetitions what assures identities of all kinds? At what "point"—at what limit of sense perception—does excess become counterproductive? In what way is repetition essential to the structuring of expectation? In what way is expectation necessary for the structuring of coherence? How immediate is "immediate musical perception"? These are but a few of the many questions raised by North, reflecting his overall concern with the listener.

Some of the analogies he drew among the arts are of equal interest. North maintained, for example, that, all the arts require "distance," as in painting, in order to grasp the "whole." What he meant by "distance," when translated into the individual arts, is not fully explicated. Still, he conceived some kind of "whole" that is not the simple addition of its parts. That he had a concept of this kind in mind comes also to the fore in his criticism of the "madrigalistic" treatment of the text-music relationship, in which "weave" and "fabric" obscure the garment, for no character is to be derived from *spot* behavior, as North, himself, learned from Aristotle. Moreover, the imitation of sound by sound, so characteristic of madrigalism, fails to convey the intended import. There are many more examples of this kind in North's writings, concerns which reverberate in the deliberations of the thinkers we have closely examined.

More than a theorist, North may be viewed as a conscientious critic. Weber (1992) justifiably characterizes North as one who prefigured Charles Burney in developing the idea of "musical classics." Indeed, the need to safeguard cultural achievements loomed large in the setting of what North called "the collapse of the republic of the seventeenth century gentlemen players," referring to the collapse of a cultural elite in the face of spreading commercialism and its attendant values. While his rhetoric, as well as his political views, may sound conservative, he was nevertheless prepared to forgo the old custodians of culture, including their social positions and functions, as far as art was concerned. Having chosen to focus on the arts themselves, he sought to elicit, from them and through them, their inherent and intrinsic characteristics. Like Shaftesbury and other well wishers at the time, North hoped to serve society at large, including those who were unaware of their spiritual needs. Divorcing artistic activity from its traditionally sanctioned settings may be construed as a sort of "secularization," yet North—the conservative—went a step further, for he opened the field for additional "players" not only to play but to fathom as well the rules of the game.

The Place of Art in a Prosperous Society

"Fathoming the rules of the game" entailed no less than a serious exploration of all that pertains to second-order knowledge; as such, it also covered the search for a clear distinction between first- and second-order knowledge. Elevating the discussion of culture to that level, constituted, in and of itself, a major change in the deliberations on art. Jacob and Harris, each in his way, represented this enhancement in the debate. Though, to the best of our knowledge, they did not converse with each other, both men participated in this momentous discourse. Our decision to join them together, despite their different backgrounds (Jacob was a poet, playwright and translator; Harris was a philosopher trained in systematic thought), stems from the fact that they shared both premises and insights.

In their view of art, the shared point of departure was a concern with well-being rather than "mere being" (to use Harris's apt phrase) differentiating between spiritual and practical orientations in human activities. "If *Well-being* be clearly preferable to *Mere-being*," the arts of elegance (i.e., music, painting, and poetry), maintains Harris, "may have reason perhaps to aspire even to a *Superiority*." Shaftesbury, as we observed, considered the arts a "luxury," resulting from conditions of a well-to-do society. Jacob and Harris took these conditions for granted, *their* cultural sphere already included inquires into the *epistemic* conditions of the various artistic media. By maintaining that the arts bestow form on function (social or otherwise), Jacob seems to imply that it is through artistic "products" and their history that one may learn what culture is all about. Though Harris introduced a distinction between the practical and the artistic into Aristotle's conception of Art, he used Aristotelian frames of thought to support this distinction. Aristotle's "four causes" are exploited by Harris to substantiate the "bringing about" of the culture with which he was concerned (*AM*: 113-5). Since intentions, volition, and consciousness loom large in all that pertains to culture, they highlight the "secondary" epistemological status of culture. All the central factors of culture are largely anchored, according to Harris, in habits of thought and acquired behavior, which supply their own rationale in their effort to realize some "absent good." Art, thus, is both circumscribed by culture while contributing to its furtherance.

The Ancients, according to Jacob and Harris, consulted Nature, they copied its Perfections, and even tried to improve upon it. Their attempts were not confined only to the arts, but affected governance, values, and manners as well. By contrast, the conveniences of life and leisure brought about by the mercantile society, despite its improvements and many blessed inventions, gave rise to "distinctions" by acquisitions, rather than by merit and value. The effects of "luxury" and "ease," which North talked about, are echoed by Jacob and Harris, who perceived them as prejudicial concerning perfections not only in the arts but in all walks of life, hindering also those who truly wish to follow them. "Well-being," as defined by Jacob and Harris, applies not only to the few, but constitutes the hallmark of the culture of an entire society.

While Jacob and Harris agreed that certain conditions might be more conducive than others to the production of culture, they, unlike Shaftesbury, no longer took the benevolence of man for granted. Ostensibly positive conditions, thus, may have undesirable consequences: "luxury" and "ease" do not only promote study, but may also cause its neglect, by shifting the emphasis from the acquisition of knowledge to that of material goods. To assure positive results necessitates a clear understanding of all of the factors involved in culture, including the powers they exert in society. Hence the reference to the Ancients, because their attempts embraced not only the arts, but also governance, values, and manners as well.

Inviting an enlightened investigation that aims to persuade via rigor of thought, rather than solicit new recruits to wishful thinking of sorts, furthers the process of secularization, for nothing remains sanctioned, viewed unto itself. Harris's employment of the Aristotelian four causes (discussed in *AM*: 113) revealed, for example, that the tendency towards social democratization does not necessarily ensure the kind of spiritual excellence from which a whole society may benefit. Shaftesbury's equation, that to be natural means to be civilized, was now broken; norms, it was now realized, are not derived but made; moreover, they change, according to different desiderata. The laws accompanying such processes have to be unveiled, alas, for what they are.

Between Qualities and Imagery

Assisted by an empiricist orientation, the focus on "secondary" elements in Jacob and Harris led precisely in that direction. Following Locke, Harris regarded knowledge as primarily sensual, considering ideas as derivative; it is sense data, stimulated by objects, which create the idea of the object. Yet the data eliciting the ideas, maintained Harris, are not only those which are considered "natural," that is, directly caused by the object to which they refer; they need not even belong to the same category to which the object belongs. Thus, through sight and hearing, through the perception of motion, sound, color, and figure which they enable, art manages to imitate either "Parts or Affections of this *natural World*, or else the Passions, Energies, and other Affections of *Minds*" (*AM*: 119). Moreover, being constricted by its own medium, each artistic domain strives to overcome its own confines. Painting, for example, tries to depict motion through static lines; poetry tries to imitate reality, as experienced, through arbitrary signs; music tries to elicit affects, in the mind, through sound and motion. Though not fully formulated, such observations clearly opened up a whole new field of investigation. Tracing the relation between line and movement, sound and gesture, brushstrokes and form, entailed no less than an investigation into the relation between sense formation and its predication. Harris, himself, provided ample insights in this direction, paving the way for excursions into the nature of artistic illusion.

While aware that the media thus aspire to overcome their limits, Harris was also aware that the media "specialize" in their choice of subject matter. Painting, accordingly, should adhere to *"actions and events,* whose *integrity* or *wholeness* depends upon a *short and self-evident* succession of incidents, or if the succession be extended, then (it should be extended to) *such actions* at least, whose *incidents are all along, during that succession, similar.*" Or, continues Harris, painting should adhere to "all *actions* which being qualified as *above,* open themselves into a *large* variety of circumstances, *concurring all in the same point of time"* (*AM*: 125). Harris's list is large, and could be further enlarged. It is not the degree of its exhaustiveness, however, which matters, but that it reveals awareness of the "simultaneity" of sight and its "durational" aspects from a perceptual/conceptual point of view.

Yet the possibility of free play between the perceptual and the conceptual in painting is limited, Jacob tells us. "To reduce the *intire Image*, which three or four Verses (of Virgil) convey to the Mind, to *Light* and *Shadow*, is impossible," he writes. "What a Complexity of beautiful *Images* are here charmed up into the Mind, as it were, by the *Magic* of a few Words? It is in the *brightest Ideas* that *Virgil* thus represents the *Transfiguration*, [of the figure he is portraying]" (*AM*: 88). While there is a vast difference between images evoked in the mind and those realized on canvas, Jacob's "transfiguration" renders nonetheless, through its *successive* images, a kind of "simultaneity"—the depiction of a *character*, a *mood*, or even the *nature* of a given process. While all of this anticipates Lessing's theory, developed in his *Laocoon,*[16] Jacob's and Harris's ideas, jointly, lead even further. For example, by allying painting and music epistemologically (on behalf of the "universal languages" which they employ) and poetry and music ontologically (on behalf of the "many copies being equally perfect with the *Original*"), Jacob anticipates cross-cultural studies in search of universals in art, and Nelson Goodman's division of the arts into autographic and allographic arts respectively.[17]

An important concept emerges from Jacob's and Harris's contemplations regarding the interaction of media and ideas. Twining will call it "obviousness"—a concept which stands for a *kind* of relation between reference and referent. "It must be confessed," writes Harris, "that-in as much as musical imitations, tho' *natural*, aspire not to raise the *same* ideas, but only ideas *similar* and analogous; while poetic imitation, tho' *artificial,* raises ideas the very *same*—in as much as the *definite* and *certain* is ever preferable to the *indefinite* and *uncertain*...it will follow *from hence* that-even in subjects the best adapted to musical imitation, the imitation of poetry will be still more excellent" (*AM*: 134). Though Harris still viewed referential relations in terms of the dictates of the mimetic theory, he no longer entertained the *directness* of perception, for not all which seems direct can be so recognized, and that which can, relies on previous learning and habituation. While Hutcheson viewed perception, at least in art, as related to levels of association and combinations, Harris and Jacob contributed to the awareness that cognition involves perceptual manipulations which, in part, depend on habits of mind. Though still unaware that "immediacy" is granted through prevailing "modes of representation" (to quote Goodman), Harris's and Jacob's conceptions seems to lead in that direction (*AM*: 134).

Harmony and Perfection Reconsidered

Harris, however, was not aware of the fact that the exercises of the mind, which he was talking about, involve the relationship between particles and wholes, impressions and patterns, successions and order. Harris's shortcomings, in this respect, stem from his overadherence to the concept of mimesis which, to begin with, fostered his attempt to rank the arts according to their success in rendering "reality." The "*accuracy of the Imitation*, and the *Merit of the Subject imitated*," are the final criteria for artistic achievement, he insisted. Adhering to the old verisimilitude conception, Harris overlooked forms, patterns, orders—"totalities" which yield meaning—not realizing that the indivisible work of art—Langer's "Art Symbol" (see chapter 4, fn. 9, and fn. 19 below)—is constituted from particles which, at one and the same time, it also governs.

Jacob understood it. Jacob's point of departure was harmony. He considered harmony *the* "constitutional cause" of art—a "congeniality" which presides over its constructedness. In Jacob's thought, a classical approach was combined with an empirical bent to yield a kind of "harmoniousness" which *transcends* reality. Jacob was also attentive to the different "musics" of the arts, to the *harmonious* combinations brought about by each of them separately. All of this became possible since Jacob ascribed "harmony" to the mind, without loading the latter with preestablished ideational frameworks. It seems as though he understood that harmony emerges from the "tendency" of the whole, which is already contained in its parts. For Jacob, "perfection" is media-related, and can never be defined in terms of mathematical relations, as Hutcheson maintained. Nor can art be estimated via external reality. Art constitutes an autonomous field, which tries to attain perfections in its multiple realizations. More than emulating nature, art may resort to successful artistic models for guidance. Hence the use of art to explain Art.

It is no wonder, then, that Jacob should have stressed that perfection is related to *technical* abilities, which enable art to cope with artistic objectives. The advance in technical abilities, however, may be related, in no small measure, to "prosperous circumstances," that is, conducive economic and cultural conditions. Poetry, for example, "required so much more *Learning*, and so many more *Arts* to bring it to *Perfection*, that it was obliged to wait for the *Revival* of all of the *Parts* of *Learning* before it could make any Figure" (see *AM*: 94). It is on this ground that Jacob meets Harris: Artistic objectives remold artistic procedures that enable the rendition of new perfections. The history of music in the West beautifully illustrates the dependence of new perfections on the cumulative interaction between theory and desiderata.

For Jacob, harmony also accounts for the *success* of works of art. Harmony, he specifies, needs a focus, a vantage point from which to perceive the entire configuration. One can benefit from artistic models, which succeeded in creating artistic coherence, more than one can learn from nature, we are told. Perfection, however, also calls for "consistency," achieved by meticulous observations of that which is being depicted. Interestingly, since harmony is concerned with totalities, it allows for a degree of incompleteness, once the tendency for completion has already been created. Deletions and omissions are preferable to additions, economy is superior to redundancy, Jacob tells us. Though he refused to draw specific stylistic conclusions on the basis of his observation, he did call attention to the role of imagination in the consummation of the "absent parts" of works of art.

Jacob, thus, adds "imagination"—much discussed at the time—to judgment and cultivation; imagination *must* submit to judgment, which *must* submit to cultivation. Coherence does not result from any one of these factors, but emerges from the balance created among them. While imagination may seem to provide the basic configurations of art, it is, in fact, conditioned by that which exists (i.e., culture) and oriented to that which is missing (again, culture). An appreciation of the products of the imagination is inconceivable, however, without judgment, itself hovering between cognitive constraints and cultural values.

Music, as expected, looms large with Jacob. Since music, as a symbolic art form, creates its own building blocks, it may be controlled and manipulated more easily, according to Jacob. The principle of "musical harmony"—paradigmatic to art forms since the days of Pythagoras—is thereby given a new interpretation: it is the *artificiality* of music which is so appropriate to harmonious organization. Harris, by contrast, ranked music below the other arts, on the basis of its imitative power; it lacks the power to *specify* that which it imitates, he said. This lack of specification, argued Harris, stems from the fact that music imitates only through sound and motion, and "there are not *many* motions either in the *animal* or in the *inanimate* world, which are *exclusively peculiar* even to any *species* and scarcely any to an *individual*." Music, thus, *"derives its efficacy from* another *source*," from its power to excite affections and dispositions in the mind. Though Harris was satisfied with the cliche about the "natural" connection between music and the passions, he nevertheless added to it an interesting notion. Not unlike Hume, and perhaps directly influenced by him, Harris presumed that there is "a *reciprocal operation* between our *affections*, and our *ideas*; so that, by a sort of *natural sympathy*, certain *ideas* necessarily tend to raise in us certain *affections*; and those *affections*, by a sort of counter-operation, to raise the *same ideas*" (*AM*: 142). This reciprocity, Harris claimed, lies at the base of the vital connection between music and poetry; music supplies, more immediately, the appropriate state of mind, by which the ideas, introduced by poetry, are more easily and willingly perceived.

Jacob saw this reciprocity as stemming from the uniqueness of each of the media: "Music," he wrote, "is a great and very sudden *Mover* and *Refiner* of the *Passions*; its *Operations* are *Intense*; but as it consists only of *Sounds*, to which no other *Ideas* are annexed, its *Impressions* are soon and easily *defac'd*" (*AM*: 89). Here, Jacob hints at an important Humeian distinction, that between impressions and ideas, which will become very central to the understanding of the nature of the "successive" arts, that is, music and poetry. Whether he was fully aware of Hume's distinction or not, he certainly understood that the perceptual nature of music is directly related to its efficacy, on the one hand, and to its lack of conceptual content, on the other.

Though Harris did not delve as deeply as Jacob into these matters, he understood that the reciprocal "sympathy," to which he called our attention, makes even the "non-plausibility" of opera plausible. In fact, Harris's reciprocity theory gains substantial support, as we shall see, from contemporary theories of affect and metaphor. For the present, however, it is noteworthy that Jacob and Harris opened the discussion concerning the "give and take" among the arts which stands to enrich their effect and import.[18]

This important insight notwithstanding, Harris's disregard for the symbolic capacity of music is not surprising, given his notion of mimesis. Still, having recognized the centrality of "compact" agreed-upon symbols in poetry, one wonders what stopped him from applying the same to music. Altogether, his underestimation of coherence made him overlook the symbolic status of the work of art as a whole, even in poetry. Harris invariably treats the symbolism of language in terms of similitude: "The *sentiments* in *real life* are only known by men's *discourse*. And the *characters, manners,* and *passions* of men being the *prompters* to what they *say*," writes Harris (Vol. II: 139). Poetry, which imitates these sentiments and characters through "discourse," thus perfectly complies with its subject matter. However, the "Art Symbol," as Suzanne Langer calls it, should never be confused with the symbolic meaning of its constituent parts.[19]

Taken together, Jacob and Harris well represent the transition from a Platonistic orientation towards an Aristotelian point of view. The imitative arts seem to have regained an important standing, as the notion of verisimilitude yielded to the way it is brought about, that is, to the *making* that it involves. We are fully aware that, with Aristotle, art as imitation connected art to the *real* world, redeeming it from the deceptive standing that it had for Plato; but even "appearances" involve some *doing*! Aristotle's understanding that imitation is natural to man, and is crucial to his cognitive development, already connected man to man, turning knowledge into a community-anchored activity. Jacob and Harris were, of course, aware of all that, but they went beyond it. Having adopted the empiricist notion regarding the origin of ideas, they even turned that which still seemed primary, in Aristotle, into second-order knowledge. New distinctions emerged from these changes, including the distinction between intentional, cumulative knowledge and a more immediate, though not "primal" kind of knowledge. The deliberations of art, as exemplified by Jacob and Harris, fall in that orbit: While cumulative and intentional knowledge highlights the role of culture in creating Culture, the "immediate" directs the attention to the processing of the mind via the senses, highlighting the notion of illusion. Paradoxically, this down to earth approach stresses what does not yet exist, and posits that which exists as mental constructions. Creative imagination connects the two, hence its importance in these and similar treatises of those years. Based on these writings, it becomes clear that long before Coleridge it was already understood that imagination is no mere fancy, but a mental power circumscribed by a combinatorics of images (in the spirit of Jacob's harmony), themselves related to a reshuffling of causes and effects concerning sensations and their referents (in the spirit of Harris's "media"). The "reshuffling" of causes in relationship to desired effects is what will concern our subsequent writers.

Comparisons and Correspondences—
Between the Moving and the Fixed: Avison, Beattie, and Webb

Effecting Musical Affects

If Jacob and Harris are identified with writers that helped establish the "modern system of the arts," as Kristeller called it, Avison,

Beattie, and Webb introduced issues which challenged the very system (which they had already come to take for granted). Whereas their thinking incorporates all that we have discussed above, they bestowed a new order on previous distinctions and the relationship among them.

By the time they were active, Hume's revolutionary theory concerning the relationship between cause and effect had already made inroads into philosophical thought. Hume, as is well known, reduced even scientific causes to habits of thought, thereby enlarging the territory of the "made" at the expense of the "given." Skeptical as Hume's theory may appear, it invited the investigation of all kinds of "associations" that had not been tampered with previously.

On the face of it, Beattie seems to belong to the group that rejected Hume's initial premises in favor of a search for more substantial truth. In his investigation of the arts, however, he seems to have deviated from "the truth as given," to the truism about man who fabricates his worlds. A follower of Reid and the common-sense school (see *AM* introduction to Beattie), Beattie was not interested in the *philosophical* position of "common sense" as far as art was concerned; he treated it as a *sine qua non* for all sense making, no matter how epistemologically established. According to Beattie, it is upon shared knowledge that artistic exercises are based, and shared knowledge is habitual. Looked at this way, art, for Beattie, is not unlike scientific theory; both are man-made constructions, attempting to differentiate between essentials and variables. As far as art is concerned, the dicta of the media pertain to essentials, whereas styles pertain to the variables. Art performs an important psychological and cultural function, but aesthetic theories are raised to the position of science. The equation of science and aesthetic theories, introduced by Beattie, is premised on the unity of the mind, the products of which bear the stamp of rationality. It is rationality which is now related to naturalness, while mimesis is related to habituation. Not only does such a view eliminate Baumgarten's lower cognitive faculty, but it does away with the multiplicity of faculties introduced by Hutcheson. Beattie overlooked neither mental propensities nor individual talents, but insisted on their interactivity. By following Reid, Beattie did not ignore the "musical ear," and considered it but one of the factors that contribute to the comprehension and the making of music.

Avison deviated from imitation, in like fashion, by introducing expression as a central issue. Of course, he was not the first to deal with expression in music, nor was he the first in our group to consider music an expressive, rather than an imitative, art. Harris, as we saw, was the first to so treat music's referential capacity. But, while Harris was still content to view the relation between music and expression as "natural," Avison, the theorist-musician, wished to demystify that belief, through some kind of concretization. His understanding of expression was novel, for it contained both the effect created by the work of art *as a whole* and the response of the listener to the combinations of melody and harmony that had not yet been specified. Thus, Avison's expression deals with the effect of the work of art and with the effect to that effect. The latter, termed by Webb *affect*, bridges old and new works of art through what the listener has learned to recognize. It is in this sense that Avison's "internal sense"—"the capacity of receiving pleasure from musical sounds"—is different from that of Hutcheson. While adhering to the concept of perfection—that which the internal sense both recognizes as well as guides—he added affect, as a mental entity with a standing of its own, as an inseparable part of the process. The investigation is, thus, directed both towards the nature of that mental entity, as well as to the perfection which elicits it, which is now reexamined in light of its consequence.

Webb enters the scene at this point. Outwardly, he seems to belong to the school that considered the relation between music and its effects as "natural." In fact, his investigation concerning the "relation between sound and sentiment" aimed at separating the habitual from the given. What is it about music, asked Webb, that makes it particularly compatible with sentiment? From Hutcheson through Beattie, this question was dealt with under the rubric of "sympathy," itself resting on an additional concept—that of resemblance. It is not that Webb forwent resemblances, but, unlike his predecessors, he was aware of the fact that a theory of resemblance must rest on some kind of commensurability. His question reveals that he was aware of the fact that sound and sentiment are in themselves incommensurable, calling for the unveiling of a factor that would provide a proper common denominator.

Even more than Avison, Webb was bound to deal not only with the dictates of the medium, but also with the characteristics of the emotions. What was implicit in Jacob now turns into a full explication: music, by definition, creates a succession of *impressions*, while poetry creates a succession of *ideas*; thus, music and poetry represent two different kinds of cognitive processes and affect bridges the two. Affect is explicitly brought to the fore as a mental entity, which Webb examines at length. Webb enlists the differentiations introduced by Hume for various cognitive activities, such as impressions, ideas, and affects, in his reexamination of the relation between mental causes and their effects. Thus, Hume, as it were, helps Hume. Webb, then, aimed to provide us with a full-fledged scientific theory dealing with all of these matters. If Beattie sensitized us to the relationship between aesthetic and scientific theory, Webb, by providing one, directed our attention to the "artistic," that is, to the creative aspect of scientific thought. It is no surprise, therefore, that with Webb the critic should assume the status of a creator (a thought already entertained by Shaftesbury), who fathoms the *lucido ordo*, which, to begin with, supplies the reference for artistic creation.[20] For Beattie, to be natural meant to be rational; for Webb to be natural meant to be creative within the constraints of systems. From all that was said, it should be clear that aesthetic theories underwent a process of increased "scientification," through a discourse which both elucidated the issues and instilled greater rigor in their treatment.

Comparisons via Functions

Given these introductory notes about our three writers, let us have a closer look at each of them, for additional points they raised to support their major claims. Of the three, Avison was the least philosophically oriented, yet his influence was enormous. A theorist and musician, he became a musical authority for later thinkers, regarding all that pertained to music aesthetics. Like our previous writers, he, too, was concerned with the public good and viewed music as serving this aim. However, unlike Shaftesbury, he did not focus on the benevolence of man, but was concerned with the benevolence of music, that is, with the positive effects which music exerts on man. Perfection in art, accordingly, can only cause pleasure, which is itself a positive social attribute. Avison asked his readers to engage in a mental exercise in which they ask themselves if they had ever experienced malevolence while listening to a successful composition. While his rhetoric in this connection is still ambiguous, for it does not completely separate semantic content, structural achievement, and reactions to the work of art, he nevertheless viewed the persuasiveness of artistic construction as a prerequisite of artistic experience.

Since expression, for Avison, related both to effect as well as to affect, it is understandable why he should have focussed on the composer, for it is the composer's art that accounts for the "powers" of music. The composer, after all, is in charge of the overall design, which dictates the hierarchical relations among parts, and determines their relative standing in the composition as a whole. It is that whole, brought about by the composer, which contains its import. The affective response of the listener is to that totality, *as it unfolds*. Indeed, Avison viewed the listener as an active participant to the decisions made by the composer, who expects the listener to follow the musical transformations which contain *built-in* expectations, as well as resolutions. Whereas particular combinations of melody and harmony may be novel, they rest nonetheless, Avison implied, on a referential probability system. Avison was ready to bestow on Rameau the recognition he deserved for having made the workings of the system salient. Within that system, Avison viewed melody as primarily "inventive," while he viewed harmony as a technique. Though both call for imagination, as well as discipline, melody is, nonetheless, according to Avison, freer in its creative aspects. Avison was far from settling the debate between Rameau and Rousseau; still, his way of juxtaposing melody and harmony made it amply clear that both operate within the same overriding tonal system. As is well known, Rameau did not ignore melody, he only considered it derivative from chordal progression. Such a view, however, overlooked the fact that melody relates to the *physiognomy* of a whole and not to the *points* on its curve. Because of the totality associated with melody, Avison, like Rousseau, viewed melody as central to the overall design of musical compositions.

In order to clarify what he had in mind, Avison engaged in a comparison among the arts, utilizing what he assumed his readers already knew. This ostensibly didactic procedure reveals Avison's understanding of analogies. Thus, the similarities to which he pointed concern functions, rather than the artistic constituents enabling such functions. For example, the function of color, in painting, is analogous to that of harmony in music, and that of space to that of sonoric layers. The various functions seem to provide new possibilities with which to conceptualize musical phenomena aesthetically. Even if the similarities he suggested are metaphorical, that does not mean that they are void of insight; as all good metaphors do, they further our knowledge. Note that here, too, Avison was concerned with effect—one of the basic roots of metaphor, which may be related to some kind of synaesthesia. Yet by exposing new relations in the so-called "target domain"—music, in this case, he went a step further.

In fact, the entire development that we have been discussing gradually led towards the dominance of *ut musica poesis*, it went from the immediate to the less immediate, from the obvious to the less obvious, attempting to clarify the *perceptual* basis of art. If Webb could more succinctly deal with "impressions," it is because people like Avison used literal images to make their point. Avison, of course, was also concerned with the correspondences between music and poetry; he could hardly overlook it as a musician. Yet, it was painting that helped him to elucidate music's constructive aspects, while poetry enabled him more readily to deal with modes of symbolization. Basic "correspondences," however, are what he was really after, trying to avoid the pitfall of simple and obvious relations among the arts. Following Harris, he was aware of the fact that poetry *talks* about affects whereas music *excites* them through the associative import of the composition as a whole. Within that whole he also allowed for imitation, which as far as music was concerned, he considered inartistic. Accordingly, he placed imitations in the accompaniment, leaving the expressive to the voice. Avison also sensitized us to the role of conventional musical coinage, by means of which composers facilitate communication. While singling out the different modes of symbolization which music engages in (an issue that will occupy those influenced by him), Avison left the symbolic function of the composition as a whole less clearly defined.

Expectations Temporalized

Beattie addressed each and every one of the points raised by Avison, but instilled rigor into the discussion by anchoring it in new cognitive considerations. Subjecting imitation to habituation, in fact, reveals that he was dealing with systems of expectations, rather than with specific symbolic meanings. Only systems that comprise built-in expectations can be enlarged to include "the probable." It is the probable that enables the creation of possible worlds, as we have learned from the Leibnizian tradition; however, possible worlds rest on the consistency within the fictive, Beattie tells us. Thus, while the fictive is invariably wedded to the imagination, its communicative aspects derive from shared knowledge. Art, Beattie concludes, is therefore a kind of "exemplification," which forms intensifications of sorts through patterns of perfection. Such intensifications can only be formed by imagination, regulated by knowledge. Hence, reality is but a point of departure and constitutes a frame of reference within which Culture assists in producing culture.

Dealing with music, Beattie was not perturbed by its lack of imitative powers as was Harris, nor did he feel compelled to make a special case for music, as did Avison, since his theory was all encompassing and equally relevant to each and every one of the arts. If at times music seems to carry specific meanings, it is through conventionally established associations between certain musical configurations and the meanings that are attributed to them. As a true iconoclast, Beattie saw the sadness of the minor mode as no less conventional than the relationship between organ music and religious sentiments. Such associations, however, are not easily changed, despite their arbitrariness, nor do they lessen the affective potency of the music. In his attempt to trace the "roots of reference," Beattie managed to show how intricately intertwined sense and structure are in works of art.

In the final analysis, what is important to works of art, according to Beattie, is the degree of their persuasiveness, rather than their attempt to achieve ideal types. Beattie's concept of the probable, as we have seen, is intimately related to the notion of the consistent within the fictive, rendering a kind of credibility that guarantees persuasiveness. Extremities, in whatever direction, should be avoided, because of the difficulty they raise for the unity of the design. Excellence and artistic achievement are related to the proper balance created between simplicity and intricacy; simplicity must be able to subsume intricacies in ways that preserve the perceptual unity of the whole. Thus Beattie exemplifies his own understanding of "unity amidst variety" through rhythm, subsumed by the simplicity of meter, which, in turn, is unified by the beat. Though maximal variety and intricacies are to be found in rhythm, the temporal dimension in art presents the problem of retention, affecting the ability to reassess that which transpired. Since mental limitation constrains the successive arts, works of literature and music, Beattie tells us, have each their optimal lengths; moreover, perception dependent on linearity is different from that of gestalts. While Beattie did not elaborate upon the latter, he revealed an awareness of the problems created by the former, in which the notion of "making" takes into account mental constraints that the making of sense involves.

Music as the Art of Succession of Impressions

Beattie only touched on the problem of linearity, while Webb's main concern, as we have seen, was with successions and the way they are perceived. Each of our writers went in a predicted direction. Given his commitment as a musician, Avison was primarily concerned with the constructedness of music, hence the correspondence which he "found" between music and painting. Beattie, a poet-philosopher, focused on what he considered "necessary and sufficient" for the communicability of that which *lacks* truth-value, from a philosophical and historical point of view. Webb, by contrast, may be annexed to the English school of metrists, which through the seventeenth and eighteenth centuries was concerned with effective arrangements of "movement" in poetry.[21] It is movement which led Webb to connect poetry and music through a *metaphorical* exchange, best expressed in his famous statement, "music borrows sentiments from poetry, and lends her movements" (*AM*: 316). It is interesting that Webb should talk about movement and not about motion, the way Harris and his followers did. Motion for Harris stood equally for imitation by musical movement, that is, metaphorical movement, and movement of objects in space, creating an ambiguity which he seemed aware of. Indeed, not differentiating between the two kinds of movement created difficulties as to the identity of that which is moved. Avison and Beattie, as may be recalled, tried to rid themselves of such comparisons, paving the way for a better understanding of that which "moves" in art. Webb seems to have benefitted from such discussions. More than his contemporaries, he understood that "*unité de melodie*" (a concept elaborated by

Rousseau) comprises gestalt as well as linearity. The latter connects the vertical and the horizontal, melody and harmony, while the former creates mental images. Condillac, too, in his discussion of gestures, seems to have pointed in the same direction. Both men understood that within the unmediated wholes there are different crystallizations of perceptual layers. By insisting that music is a *succession* of impressions, but that these impressions are meaningless *out* of context, Webb, however, went one step further, anticipating Bergson's understanding of movement as an immediate and unmediated construction of consciousness.

While the concern with movement neither began nor stopped with Webb, he succeeded in clinching several aspects in a way that is unique even from the perspective of our own times. Moreover, movement was only one of the issues that Webb addressed in his attempt to understand the relations between sound and sentiment. The overall problem with respect to what "moves" in music led him to deal with (1) the elements which are essential in the perception of music; (2) the mechanisms characteristic of affect, both with regard to the feeling of affect, as well as the procedure whereby affect is identified; (3) the compatibility between the labeling of affects and their experience; and (4) the nature, as well as the epistemological standing of the synthesis of all of these factors.

Commensurating Sound and Sentiment

As may be recalled, Ficino was aware of the relatedness of these factors quite early; he did not distinguish, however, between the perceptual and the physical in musical movement, and therefore turned the whole relationship between that which has movement and that which is moved into a simple causal relationship. By concretizing "matters" the way he did, he shrouded the whole issue of affects, their content as well as their identification, in Platonic mystery. By contrast, Mersenne, who delved into the nature of sound, was able to separate the physical from the mental, as far as movement is concerned. In fact, his isomorphic theory rested on this separation. However, granting the "made" in music more recognition than it had ever received before, he bestowed naturalness on the isomorphic theory, thereby introducing contradictions in his own explication. Still, unlike Galilei, whom he admired for having tried to make music more communicative, he understood that the association is not between entities of the same domain, that is, intonation in music and intonation in language, but between music as a whole and affect as experienced. Descartes understood Mersenne's intent, and although he viewed it as basically untenable, by delving into the nature of experiencing affect and the associations it evokes, he contributed to its clarification. According to Descartes, experiencing passions, as we have seen, is made possible by the physiological and neurological "givens." Moreover, the ability to *create* associations, like the associative ability itself, is, likewise, naturally given. The *contents* poured into associations are, however, arbitrary though historically explainable. Though he understood that the multiplicity of emotions rests on a few basic dimensions that he considered innate (Descartes 1985 §52 p.349), he did not explicate the standing of the multiple cognitive contents of emotions in relationship to these basic dimensions.

It is surprising that Mattheson, a hundred years later, should have overlooked the achievement of both Mersenne and Descartes, of reuniting the physical and the mental in a simple causal process. Nonetheless, Mattheson deserves credit for having understood Descartes's emphasis on the cognitive aspects of affect. He also distinguished, as Kivy (1980: 41-45) points out, between semantic contents and affective response. Kivy is right in accusing Webb for having deemphasized this separation, but this does not justify his having overlooked the more impressive and original separations that Webb *did* introduce. Following Descartes, Webb explicitly distinguished between the mechanism which transmits the behavioral codes of emotions and the behavior itself, stating that the former is neurological while the latter is psychological. Emotions felt do not guarantee preestablished identifications; were this the case, Webb implied, artistic manipulations would be impossible. Movement is part of psychology, hence the commensurability of emotions and music and the indetermination of both. This, however, did not satisfy Webb, the critic. As if taking the hints of his predecessors regarding the basic dimensions of emotions, Webb created a kind of mental space in which he was able to locate both primary passions, leaving room for nuances of sentiments. As we lengthily explicate in *Arts in Mind*, this enabled him to accommodate a whole array of labeled and "semantically differentiated" emotions in a way that explained the interaction between them and the space to which they belong. Webb made clear, thereby, that the artistic world superimposes a semantic space on the emotional one. This, he argued, would have been impossible without affect that bridges the two.[22]

It is surprising how modern Webb sounds, especially if we bear in mind the current attempts to account for the basic dimensions of emotions, the new developments in theories regarding affect in general, and the interesting connections between affect, metaphor and creativity, in particular. Whereas in theory the basic dimensions of emotions are still far from being conclusive, empirical research does prove their tenability, as far as emotional semantic spaces are concerned. How many dimensions there are, and whether they refer to scales such as "pleasantness-unpleasantness," "intensity-relaxation" or "extraversion-introversion" still remains an open question, particularly for those in search of universals (see, for example, Plutchik 1980). Theories, such as Izard's (1971), which entertains the possibility of some fundamental emotions, do not necessarily contradict the search of dimensions; their account of the developmental aspects of patterns of recognition may supply a kind of *infrastructure* for the organization of the complicated network which crystallizes later in life. Some co-ordination between the diachronic and synchronic aspects is also relevant to an overall theory of music expression. While expressive idioms are related to the determined meaning of accompanying texts, they are more general and flexible in themselves. Moreover, they tend to be associated with other expressive patterns, creating complicated symbolic configurations, as we have argued in chapter 3. Webb's analysis of the levels of affectivity in poetry—starting with consonances and vowels and ending with prosodic procedures—reveals such awareness. The same theoretical presuppositions serve to unveil both the expressive potential of what seems fixed though arbitrary, that is, lexical units (particularly monosyllables), as well as the built-in semantics of what seems free yet constrained—the poetical movement. Affect emerges as a *guiding* precept in both stages, for it bridges the moving and the fixed.

Thus, according to Webb, affect directs the search for a primordial communicative substratum, leading, in later stages, to that which enables unique expressions. Affect, argues Miall nowadays (1987), functions in certain cognitive processes as a semantically free, though identifiable agent, connecting disparate domains by throwing them into a relationship which conveys new meanings. According to Miall, this process is central to the act of creativity, illuminating both the root and routes of metaphor. The *in*commensurability of emotion and music, of music and poetry—the basic problem which precipitated Webb's theoretical attempts—turns out to be its solution: the relation between sound and sentiment involves an affective linkage which makes it possible for the aesthetic and the psychological to intertwine creatively.

From "Immediacy" to "Discrepancy"—Between "Illusive Sensations" and the Verity of " Disbelief": Twining and Smith

Imitation as an Open Concept—Complementary Views

As we have seen, the initial investigations, which dealt with art from the perspective of perception, were in dire need of conceptual frameworks that were philosophically informed. A strange mix between Platonism and empiricism was thought, at the beginning, to provide the expected guidelines. As time went on, however, the preoccupation with culture and the making of culture made the need for sound philosophical underpinning less acute. With the increased clarification of some leading concepts and issues concerning culture and constituent artistic elements, the need for a philosophical orientation rose again. Indeed, in the eighty years between Shaftesbury and Smith, the Platonic framework gave way entirely to Aristotelian thought. As we have seen, it was Harris who was the most Aristotelian of our writers, although Platonic vestiges still marked his thought, particularly with regard to the "naturalness" of affect as far as music was concerned. All such vestiges seem to have disappeared with Twining, who not only provided the scholarly world with an authoritative translation of Aristotle's *Poetics*, but also fully endorsed his thinking. The commentary that accompanied Twinning's translation is ample proof of his commitment to Aristotle's basic approach. Nonetheless, Twining was aware of the transformations in thought and culture, since ancient times, that were bound to affect conceptual frameworks and practices. Thus, for example, while analyzing concepts in relation to essence and distinctness, in Aristotelian fashion, he embedded them in philological concerns that were foreign to ancient thought.

Unlike his predecessors, Twining refused to deal with representation as a closed concept, focusing instead on its "uses." It is this unique approach, which he systematically carried through, that makes him sound almost like an analytical philosopher. It entailed a new way of conceptualization which started from the empirically given, rather than from a postulation of an ideal state. In the empirically given, "points of reference" have to be established, making it possible to deal with various cultural manifestations that are part of the same field or concern. If this sounds circular, it is because it rests on the premise that man can understand only that which he himself has created, and moreover, that which is created rests on ways of representation which have already become entrenched. Twining explains, for example, that the absence, in ancient Greece, of a well-developed picturesque style in poetry, is due to the fact that only much later did painting provide the kind of viewing that bestows unity on a sight via the *qualities* which it imparts. "The Greek poets," he says, "did not *describe* the scenery of nature in a picturesque manner, because they were not accustomed to *see* it with a painter's eye. Undoubtedly they were not blind to all the beauties of such scenes; but those beauties were not heightened to them, as they are to us, by comparison with painting—with those models of *improved* and *selected* nature, which it is the business of the landscape-painter to exhibit. They had no Thomsons, because they had no Claudes. Indeed, the influence of painting, in this respect, not only on Poetry, but on the *general taste* for the visible beauties of rural nature, seems obvious and indisputable."[23]

Smith, in his investigations, was also concerned, like Twining, with the empirically given as a starting point. However, as a moral philosopher, he was primarily concerned with the social-psychological factors that regulate and guarantee modes of communication among individuals, contributing to the improvement of society as a whole. Unlike Twining, he was in need of an underlying assumption, concerning the nature of man that would be compatible with these integrative objectives. For his part, Smith tried to bridge between Hobbes and Shaftesbury; from the former he adopted the self-interest of the individual and from the latter the belief that the individual, even as a free agent, contributes to the public good through his self-motivated behavior.

With Smith, the Shaftesburian concept of benevolence gave way to the concept of sympathy, which implies what was later termed empathy, namely, not only the awareness of others, but the ability to identify with them, as well (Smith 1976). Self-consciousness, in other words, involves the awareness of others; a "private" sense of the good is a contradiction in terms. This socially anchored concept also helps Smith to explain different modes of communication, as well as his aesthetic theory.

Moral sentiments, however, operate at different levels of intensity, depending on the "distance" of the external stimuli that impinge upon them. The further the distance, the less the intensity. At an extreme distance, there is limited occasion for any moral sentiments to take place. Still, proper behavior rests on the ability to overcome distance; the rules of justice are conceived for that purpose. The concept of sympathy and distance, thus, turn out to be closely related. The former serves as a prerequisite of the latter, because it establishes a distinction between the self and others and the latter is a measure of the degree of the distinction between them. The psychological and the cultural *thus* intertwined guided Smith in all of his investigations, whether moral, economic or artistic (see *AM*, introduction to Smith). Applied to art, it focused on the nature of the artistic objects, that is, on their "distances" from nature and on the way they overcome that distance.

Between Obviousness and Immediacy

While both Twining and Smith tried to clarify imitation in art in terms of modes of representation, the former focused mainly on the consumer while the latter focused mainly on the producer. The two are interrelated, of course, for the producer of art makes assumptions about the consumers, counting himself as one of them. Yet the different emphases yield a kind of understanding which is of a complementary nature, rather than an overlapping one. That both Twining and Smith should have shared the concept of illusion as a key concept does not contradict the previous statement; since the mental image is, indeed, the meeting ground of the perceiver and that which is perceived.

The kinds of mental images, created through different modes of representation, is what Twining wished to unveil. "Proper imitation" is both obvious and immediate, Twining postulated. He was able to do so because he used the visual arts as paradigmatic to all artistic representations. While "obviousness" is more cognitive, related as it is to some kind of referential power, "immediacy" is more perceptual, for it is related to the recognition of the image *in* the artistic object. These two conditions explain resemblance in painting and sculpture. Twining also traced the way artistic resemblance takes place under less "favorable" circumstances. What happens in fact, asked Twining, when one of these conditions is lacking? How does one compensate for its absence, he asked, shifting the emphasis from resemblance as such, to the ways and modes by means of which resemblances are achieved.

Twining was not misled by simple analogies that may be created among various media. Words, he told us, can be both immediate—when employed onomatopoeically—as well as obvious—when used lexically. However, this kind of linguistic symbolization cannot yield an integrative image or constitute a work of art. In poetry, what is considered to be of significance lacks immediacy, which neither alliterations nor the most powerful onomatopoeia are able to compensate for. Nonetheless, good poetry creates mental images. While Twining did not entertain a theory that would provide the base for the illusive power of poetry, taking the latter for granted, he was able to reveal a process whereby the cognitive may yield imaginative perceptual qualities. Twining not only observed that the same medium may give rise to different kinds of illusions, all of which may occur in the same work of art, but also highlighted the indispensability of the integrative nature of artistic mental images.

Twining also insisted that the imitating ideas should enjoy both "distinction" and "vividness." Immediacy, thus guaranteed, makes the application of imitation in poetry almost akin to that of painting, argued Twining, explaining the basis of the "*ut pictura poesis*" tradition. Indeed, distinction is not easily achieved as far as audible and mental objects are concerned. The latter, Twining contended, reminiscent of Webb, does not admit of description at all unless approached through its various effects, whether internally or externally revealed. Complexity is, therefore, an integral condition of the description of passions, as it is (to recall Baumgarten) of poetic description in general. In accounting for one's own feelings, description becomes expression; however, it may be called imitation when it is included within a larger fictive scheme. Immediacy, as far as hearing and feeling are concerned, is, thus, achieved even more *in*directly, for here the effects or impressions are themselves vague and ambiguous.

If cognitive processes have the power to supply what is lacking in imitative language, why not assume cognitive participation in what Twining considers immediate, that is, the visual? The latter is, in fact, implied in Twining's argument, although not explicitly stated. Again we are faced with a problem that has not been resolved, even nowadays, namely, in what way is perceiving different from thinking, or, alternately, how cognitive is perception itself? By insisting that perceiving qualities in art is attained in nondirect ways, Twining made a step in the direction that has eventually been picked up by people like Arnheim, Gombrich, and others, in the field of art, and a variety of researchers and scholars who deal with the "nondirectness" of perception.

Paradigmatic, in this respect, is his approach to music's imitative power: The "power" of music constitutes Twining's starting point. Analyzing "musical effects," Twining owed much to his predecessors, especially Harris, but he had the advantage of considering their arguments within a more systematic and larger conceptual framework. Circumscribing the limits of musical imitation, Twining viewed all music that aims either at sensual effects, or at a free play of associations, as unsymbolic to begin with. Moreover, in music, which lacks explicit references, "imitation by sound" is likewise unsymbolic. Albeit that with the aid of poetry clear reference may be obtained, imitation of this kind, in music, is never of central importance, for it is the emotions "through which the great efficacy of music is revealed."

Unlike Harris and Beattie, who believed in the dichotomy between "imitation" and "expression," Twining regarded expression as a special case of imitation, that is, imitation of the emotions. Unlike the Moderns, the Ancients, Twining tells us, conceived the "power of music" precisely this way, they understood that resemblance is essential to all imitation. Though strict representation may not be achieved in music, as "expression" it should, nonetheless, involve some kind of referential relation established through resemblance. The Ancients understood, he added, that it is only the reactions, the extraneous effects, rather than the mental affects, that are liable to be imitated. Twining, likewise, attributed to the Ancients the insight that words should be applied to music, in order to supply music with the "distinctness" of the imitated passion.

Though reminiscent of Webb, Twining's conception of imitation, in terms of resemblance between work and object, is at variance with Webb's, who viewed "imitation" as a "borrowed operation." Vocal music, though imitative, is far from fulfilling the requirements of strict imitation, according to Twining, and when it is nondramatic, it refers to the imitated passion only indirectly. "Between sounds themselves and mental affection there can be no resemblance," argued Twining, "the resemblance can only be a resemblance of effect" (*AM*: 356). For Webb, such a resemblance was enough, since passion "in itself" is forever unknown. Twining might have agreed with Webb, for he himself had a similar insight with regard to poetry, yet the passions, for Twining, had a reality, even if it resisted formulation. Music, which deals with passions, is, by definition, not immediate. Music's distance from the thing to which it refers is compounded by the fact that it does not share a "sensible medium" with the effects it describes. Twining perceived that music of this sort involved deliberate calculation, of the kind totally foreign to Webb. First the music is heard, then impressions are examined in relation to what one knows to pertain to the particular passion which the words define. It is not clear at all, how, according to Twining, the comparison is handled; rightly observing that "it is not quite philosophical" to talk of such a resemblance "as being in the sounds themselves," he failed to explain, however, how resemblances may be ascribed to them altogether. Perhaps he did not bother, for his main attention focused elsewhere.

In "dramatic music" he finally found a full parallel to dramatic poetry. It is in music associated with drama that, according to Twining, the requirements of both "immediacy" and "obviousness" are fulfilled, since "the words themselves are imitated immediately, by tones, accents, inflexions, intervals and rhythmical movements, similar to those of speech" (*AM*: 359). From this theoretical point of view, the recitative, of course, is to be preferred. Viewing imitation in music as related to manners, was unfortunate; Aristotle, Twining believed, was caught in "a popular and unphilosophical argument." The *Poetics* notwithstanding, Twining tried to find in Aristotle an alternative approach, so as to conform with his own ideas. However, his interpretation of a problem, he wrongly ascribed to Aristotle, overlooked the kind of movement pertinent to music.

Pseudo-Aristotle differentiated between two kinds of motion, belonging equally "to the objects of our other senses," the one physical and the other bearing "a resemblance to human manners, both in the rhythm and in the arrangement of sound acute and grave" (*AM*: 362). The latter is, of course, metaphorical—not only with regard to its correspondent object, but in and of itself—a metaphorical movement. For Twining, however, the metaphorical level hardly existed. This negligence explains his lack of insight into the music of his time. Not explicitly associating himself with the camerata, but rather with later thinkers like Hutcheson and Rousseau, his arguments are, nonetheless, reminiscent of the former. While Purcell and Handel were among his favorites, it is the music of Pergolesi, which he preferred. His preference may be ascribed to opera buffa's realism, namely, the "verisimilitude" achieved in the *parlando* style, characterizing its protagonists. On the whole, Twining seemed unaware of the development that took place since the first decades of the seventeenth century in dramatic music away from literal expression. He did not see the operatic aria as a kind of symbol, metaphorical through and through, nor did he realize what kind of metaphor it involves, nor could he. Adhering to his own notions, he nevertheless arrived at a better understanding of the rhetorical nature of the new instrumental music. While music, for Webb, resides on a plan that extends between expression and resemblance, Twining's is located between "resemblance" and "sense data" (the realm of painting, according to Webb). Like Harris, and other thinkers associated with the *Affektenlehre*, he emphasized the natural aspects, opposing those who related certain expressive features in music to stylistic conventions. Like Beattie, for example.

Art between Illusion and Contemplation

All of these observations (inclusive of misunderstandings) go, of course, beyond the work of art, related as they are to the workings of the mind. Indeed, the preoccupation with "problem-solving" in the domain of art involved assumptions and conceptions about the workings of the mind, in both theory and practice. It was Smith who was aware of the fact that all of art is related to problem solving and said so. His whole notion concerning distance and the overcoming of distance (a reversed mirror image of Twining's "immediacy" and the required compensations to fill its absence) acknowledged the kinship between art and science. The different modes of overcoming the "discrepancies," between the original object and the imitating medium, amount to means for the achievement of ends: creating in two dimensions the illusion of three; bringing about in a static object a semblance of that which moves; prefiguring in the audible that which is unheard. Distance and artistic means, in aesthetics, function, for Smith, like distance and sympathy in ethics. Means, like sympathy, become ends in themselves; they involve the idea of contemplation. The association of means with contemplation first dawned on Smith in his *Theory of Moral Sentiments*. There, Caygill tells us, he conceived the universe itself (not unlike Hutcheson) as an immense machine for the production of happiness, emphasizing the enjoyment which consists in the contemplation of machines—of "means abstracted from the proper end" (Caygill 1989: 88). The concentration on means in themselves, Smith argued, is not only the secret principle of the development of civilization, but also the origin of the sentiment of beauty. Later, in his essay on the imitative arts, he stated that in art "the pleasing wonder of ignorance is accompanied with the still more pleasing satisfaction of science. We wonder and are amazed at the effect; and we are pleased ourselves, and happy to find that we can comprehend, in some measure, how that wonderful effect is produced" (*AM*: 389). Pointing to the double experience involved in the apprehension of art, Smith implies that cognition is embedded not only in the making, but that being cognizant of that cognition is no less essential. Since suspension of disbelief is necessary for the former, and its retention is obligatory for the latter, this "double faced" aspect of works of art should reside in their very appearances, inviting the beholder to experience both. Reality, as far as art is concerned, is a contradiction in terms, for real art never deceives as to the distance it keeps from the imitated object.

The concept of distance, the double nature of the artistic experience, and the nondeceptive nature of illusion are, thus, three aspects of one and the same conception. Judgment, accordingly, is not a mere option in perceiving art; it invariably involves the artist, as well as the consumer. Resemblance functions, for both artist and consumer as a point of reference, more than it does as an artistic end. It enters into the act of judgment which takes into account other relevant factors, like the kinds of material involved in the making, the amount of work invested, and other intentional and contextual factors. While originality looms large as an aesthetic value, beauty, as well as resemblance, are two of the many possible realizations of the aesthetic, that is, of the perception of *fitness* between means and ends. Here, the disinterestedness principle, initially formulated by Shaftesbury, reaches an interesting shift. For Shaftesbury, as discussed above, it stood for the minimization of utility in favor of spiritual gain, introducing a separation between the moral sense and that of beauty, a separation in that which initially aimed to be a *unified* theory of values. In the language of Harris, it was transformed into the distinction between "mere being" and "well being." For Smith no such distinctions exist; civilization, he believed, originated in the transition from "natural indolence" to "civilized industry," in the reorganization of the "given" according to a higher rational principle, instinctively felt. Thus, economic considerations are germane to art, not only because art works are objects on the market. Smith seems to follow Mandeville in arguing that the *active* desire for material objects underlies not only industry, but artistic objects as well. Smith implies that art shares the economic concepts of merit and value. Still, despite real and analogical connections between art and economics, it seems that Smith succeeded in developing a frame of reference that distinguishes art from other cultural phenomena. At the same time, unlike Lessing's theory, his theory of artistic media allows for *changing* artistic desiderata, provided that they obtain a proper relation among distance, illusion, and the transparency of "problem solving" *within* the contemplated objects.

"The Condition of Music"

Smith's views, regarding the nature of art and civilization, affected directly his ideas about music, as an artistic medium. In contradistinction to many a thinker of his time (including Twining), Smith did not consider the origin of music as related to the heightening of the inflections of the voice, or to other primitive communicative devices. Music, he claimed, is *the* art of leisure, its articulate power should be distinguished from those conveying "natural" meanings. Though Smith proceeded from the concrete to the abstract, from meanings that came to be associated with music to music's pure constructive elements, a reverse procedure reveals more clearly the nature of his argumentation. Of course, the reverse order amounts to history turned upside down (starting with "emancipated" instrumental music, going through more programmatic music, continuing with the aria and closing with recitative and dance). Thus, the nature of music, as an independent artistic medium, is better revealed in that which is purely instrumental, in that which has abandoned almost all relations to an exterior reality. The discrepancy principle, as it were, reaches its extreme point on the verge of its annihilation: "Why should it [music] embarrass its melody and harmony, or constrain its time and measure, by attempting an imitation which, without the accompaniment of some other art to explain and interpret its meaning, nobody is likely to understand?" Indeed, why should it, if "without any imitation, instrumental Music can produce very considerable effects?"[24] More than in any other art form, instrumental music highlights the fabrication itself, argued Smith in a rather pioneering way, raising the arguments of Avison and Beattie to a new level. The referential and syntactical rules, which music employs, enable coherence—the creation of a self-sufficiency implying modes of anticipation as well as of retention. This overall constructiveness of music, Smith maintained, is as complicated as the linguistic system, though it "signifies nothing."

It is this constructiveness which lies, according to Smith, at the heart of the perception of a piece of music, since "by means of this relation each forgoing sound seems to introduce, and as it were prepare the mind for the following." Moreover, "by its rhythmus by its time and measure, it disposes that succession of sounds into a certain arrangement, which renders the whole more easy to comprehend and remembered" (*AM*: 417). While Jones ("the Musician") still preoccupied himself with sporadic comparisons between musical and verbal segments (Jones 1784), Smith, considered time as the key dimension of both systems, and saw the metaphor of its passage applying only to the whole; only as such can music relate to the conception of a discourse in literature, in which expectation, disappointment, resolution, and the like are central. Smith even differentiated the ongoing "process"—the perceiving of music—from the finally constructed musical form, though both may be explained via the mechanism of one and the same code.

That the cumulative effect of a composition differs from the momentary perception of its parts, is a notion that will reverberate both in structural approaches to literature, as well as in cognitive treatments of music. It will turn out to be of even greater significance in the attempts to view the two arts in relation to each other (Levi-Strauss 1969; Johnson 1972; Kramer 1984). Smith's short analysis of instrumental music is, indeed, bursting with theoretical implications. His conception of system or method, for example, implies a differentiation between a general code—common to works of art, to a style, or even to the whole of an art form—and a particular code, governing a specific work. Yet, as in the visual arts, so in art in general, the one is dependent on the other. It goes without saying that Smith conceived of the musical "system" in terms more comprehensive than mere syntax, for meaning in music, he maintained, is "complete in itself." He was careful, however, not to identify the entire "system" with the organization of a single parameter, for he saw music as a construction, to which many parameters, working together, contribute.

Smith's view of instrumental music can be seen as a forerunner of Hanslick's view, preparing the way for people like Pater. In any case, for Smith (if not for Kant), the ascribing to music a kind of independent beauty, implies that the mind perceiving music does not only enjoy "a very great sensual, but a very high intellectual pleasure, not unlike that which it derives from the contemplation of a great system in any other science" (*AM*: 418). Again, the two cognitive levels of Baumgarten are united. According to Smith, another kind of cognitive unification, parallel, though not identical with the first one, takes place in the perception of music. Expression, Smith maintained, is an immediate and necessary consequence of grasping the structure of music. In Wittgenstein's version of the same idea—"Feelings accompany apprehension of a piece of music in the way they accompany the events of our life"—despite the Schopenhauerian tint, the basic conception has not changed (Wittgenstein 1980: 10e). Altogether, a new way was cleared to deal with both structure as well as expression in music. Conceiving the former meaningless, unless related to a cognitive response, it freed expression from specific contents, relating it, instead, to the psychological resonance of meaningful structures.

Smith did not deny the "semantics" of music, but insisted that music acquired it, since ancient times, from either gestural or verbal semantic attachments. Smith, thus, differed from Condillac and his followers, who viewed expressive content as *inherent* to the "language of action." While the "conventional" aspects concerning expressive music are emphasized by Smith, he does not altogether forgo some natural aspects of expression. For Smith, perhaps more than for many others, words were important for the very making of symbols. Music, accordingly, becomes an imitative art only when "married to immortal verse or even to words of any kind which have a distinct sense of meaning."

Though unimitative to begin with, music's various constituents and procedures can be manipulated so as to yield different modes of representation, assisted, if necessary, by textual labelling. Thus, it is within music's power to articulate that which enables the distinct characterization of protagonists, as well as atmosphere, in opera. Improving on Rousseau, Smith argued that music's autonomy qualifies music to render what is removed from "reality"—secondary impressions, that is, indirect expressions. Moreover, the scalar organization of musical parameters creates "orthogonal spaces," as it were, making it possible to "locate" different emotional states. At this point, Smith seems to follow Webb's "coincidence theory," connecting it to a general theory pertaining to the medium. Yet, while Webb conceived "movement" in terms of intensity and speed (the two axes constituting the suggested "model"), Smith paired "speed" to the "variety of contrast," arguing that given a "wanton and playful disposition of mind, variety of contrast is more agreeable to us than the sameness of resemblance." By contrast, in "melancholy and desponding we frequently find ourselves haunted, as it were, by some thought which constantly pursues us" (*AM*: 407). Of greater significance, however, is their different conception of movement in music. It is to Webb's credit that he saw the movement in music as a totality; Smith's identification of speed and variety with rhythm and pitch, respectively, is not only inconsistent with his own arguments, but, more important, does not make sense from a musical point of view.

Emphasizing music's special condition, Smith could, nonetheless, argue that the unconsummated elaboration of a single motif enables an aria to represent a person who is haunted by a thought or an idea which, "when he has chaced it away, immediately returns upon him" (*AM*: 398f). "Correspondence" also explains good performance in opera, like gestures correlated to time and measure. These and other modes of representation count as kinds of overcoming distances; they are measured, Smith implied, not by totally eclipsing distance but also by its retention, exploiting the full potential of the medium. The recitative, thus, is aesthetically inferior to "airy" music, argued Smith (in contradistinction to Twining). His conception of the illusion involved in the perception of artificial fruits, applies equally to music imitating natural sounds, arguing that such an imitation, when continued too long, appears to the listener as a "mere trick," a deception. The tendency of our thinkers, since Harris, to underestimate this kind of imitation, now acquires new philosophical support.

More than all of his contemporaries, Smith succeeded in granting the different genres, and the various modes of music of his time, an aesthetic status, without resorting to metaphysics to back it up. In comparison, Twining's elaborate conceptual framework scarcely enabled him to deal with the basic constituents of the medium as they came to the fore in instrumental music. Recognizing its *felt* aesthetic qualities, Twining is only able to tell us that such music "is not *imitative*, but if I may hazard the expression, merely *suggestive*." Its suggestiveness, says Twining, focusing, as expected, on the perceiver's standpoint, leaves the hearer "to the free operation of his *emotion* upon his *fancy*, and, as it were, to the free *choice* of such ideas as are, to *him*, most adapted to react upon and heighten the emotion which occasioned them" (*AM*: 357). Twining's theory, which is at ease with recitatives, madrigalistic effects and the like, breaks down in face of the "immediate-nonobviousness" of pure music. His intellectual honesty and the descriptive nature of his theory yielded, nonetheless, a notion, complementary to that of Smith, of the indeterminacy of musical import, and its powerful imaginative drift.

* * *

From all that our writers told us about music, it is clear that since Hutcheson, they had taken for granted the nondenotative nature of the medium, acknowledging, at the same time, its unique expressive power. Aware of these seemingly contradictory observations, they tried to delve into the perceptual conditions which give rise to either. Directly or indirectly, they tried to distinguish and to link three main musical aspects: (1) the interconnectedness of its particles; (2) its "impressionistic" nature (which resists reduction into "compact" symbolization); and (3) the kind of semblance created between musical and behavioral gestures. Avison, and in a more sophisticated manner, Smith, explicated the derivativeness of the first two aspects. Via the "reordering" of music's parametrial relations, they contributed to a de-mathematization of the computativeness Hutcheson ascribed to the aesthetic, advancing the Leibnizian "counting of the soul," as much as they could, to the conscious frontiers of the perceiver. New understandings of the perceptual were thus obtained, involving basic cognitive elements, such as reference and categories, combinatoric rules and hierarchical levels. The linkage between the cognitive and the perceptual *thus* exposed, threw new light on the "commensurability problem"—on the attempt to combine the "impressionistic" and the "expressionistic" dimensions of music. If contentless thought is cognitively possible, investing it with meaning cannot yield simple denotative results. The exchange (between meaning and gesture) is unique, Webb told us, for it bestows on the ideational realm the *quality* of musical impressions. This cannot be translated or paraphrased, by definition, for the supposed "realm" which music brings to the interplay does not really exist, as its various nominations—"ineffable," "inexpressible," etc.—testify. A new symbolic entity is nevertheless created, in this unique interaction of quality and label. If all good metaphors, as recent philosophers and psychologists tell us, involve unpredictable interactions, what our writers must have had in mind is, in fact, a special kind of metaphor. It is a metaphorical interaction in which the "target" (music, in this case) is basically nonliteral, refusing consummate designations.[25] The power of the same music to be differently matched, without sacrificing its general meaningfulness, is an important issue in this regard, as Webb and Twining told us. Impression implies semblance, and though its meaning is not determined, the latter is, nevertheless, affected by the structure of the appearance of that impression. Verbal metaphors are likewise unbounded, and are able to create new resemblances; the advantage of music is that the metaphorical interplay it suggests, as Twining argued, is also created in the absence of a literal "frame;" hence music's eventual role in culture as *the* free, imaginative artistic medium.

Though Smith did not explicitly discuss the metaphorical nature of musical expression, his ideas, nonetheless, acknowledged "realms" which have no literal connection with the things they imitate, which bestow aesthetic value upon metaphor. In a way, Smith's ideas allow for the possibility of viewing all art as a kind of metaphor. With regard to the relation of music and literature, at any rate, Smith's ideas carry us to metaphorical exchanges between the arts, which we deem central to the whole issue.

Most of Smith's suggestions with regard to "exchanges" are implied in his observations concerning the "limits" of the arts. Definition of "limits," we have argued, constitutes a prerequisite for any attempt to deal with metaphorical interaction. For example, if language is distinguished in its variety—in the many ideas that may serve to delineate a single subject—and music in its "repetitions" of the "same" idea, it follows that poetry may become "musicalized" and music "rhetorized." A similar idea appeared in Webb: whereas poetry moves from the general to the particular, music advances in the opposite direction. In the making of metaphors, each art enlists its own means in ways which were not yet exploited. By the time of Smith, music had already found the appropriate ways to achieve desired effects; in fact, contrasting ideas were employed as a device in the classical style to create dramatic semblance. Indeed, ideas of the kind provided by Smith were the basis for further discussion of the subject and of actual "experiments" in the centuries to come.

None of our writers explicitly discussed metaphor, yet their writings nurtured subsequent discussions that acknowledged the "power" of metaphors. What Webb had to say about the indeterminacy of the mind (with regard to the knowledge of the passions), for example, has almost direct cognitive bearing on the subject of metaphor:

> In order to treat of the passions with precision, we should determine their several modes and fix an unalienable sign on each particular feeling. To this end we should have a perfect intelligence of our own natures, and a consummate knowledge of every thing by which we can be affected: in short, we should have conceptions in all points adequate to their objects. Such knowledge would be intuitive. We should, in this case, want no comparisons of our ideas and sentiments; no illustration of one thing by its resemblance to another: thus every proposition would be reduced to a simple affirmation, the operations of the understanding would cease, and the beauties of the imagination could have no existence. Providence has judged better for us, and by limiting our powers has multiplied our enjoyments. The wisdom so conspicuous in the abridgement

> of our perceptions, appears with equal evidence in the bounds prescribed to those arts which were destined for our delight and improvement. (see *AM*: 278)

Understanding, passion, and creativity, maintained Webb, have a lot in common; they create relations among separate things, introducing novel entities to the mind. Reading Thomas Brown, who wrote some thirty years after Webb, one cannot fail to recognize a consummation of notions and ideas that were raised by Webb and his circle. Barry (1987: 124) tells us that Brown differentiated between *association* and *suggestion* (not unlike Twining), identifying the latter as that which releases the mind from the more fixed mechanism of a repetition "associated" with the former. Through this process, Brown argued, the poet can create and vary metaphors, "though there may never have been in the mind any proximity of the very images compared." Metaphors, thus understood, become hence the essence of poetry, and the active powers of the mind at large.

It still remains for historians to work out how the romantic conception of poetry (such as Wordsworth's and Coleridge's, which fully adopted these ideas) was transformed into the modern notions of metaphor, creativity, affect and their relations. Modernism, in art as well as theory, was undoubtedly pivotal in this respect. At any rate, it is not surprising to find a cognitive psychologist, of Jerome Bruner's stature, interested in connecting such notions, in his attempt to understand "the mode of knowing" that the experience of art involves. Bruner expounded the following four aspects related to the experience of art:

> The *connecting* of experience that is the reward for grasping a work of art, the *manner* in which achieving understanding of a poem or picture requires an expression of human effort, what it is that is *"moving"* about experiencing an object of beauty, and wherein lies the *generality* of that which we find beautiful. (Bruner 1965: 59)

These four aspects are obviously interrelated; the first is identified by Bruner, as the invariable search of art "to break free from well-established representation to a new, more comprehensive unity," through construction and exploitation of empty categories, that is metaphors. Joining dissimilar experiences in a compact way, and its understanding, requires, indeed, a special kind of effort. Those who are willing to undertake such an effort, tells us Bruner, are rewarded by the conversion of discordant impulses into a single satisfactory and affective image. This, in turn, affects awareness in general, we are told, impelling the beholder to seek new kinds of connections, both in art and outside art. Art, thus, not unlike science, contributes to the increased unity of our experience of reality, only that in art it is achieved via metaphorical grammar.

By now, it should be clear that our writers have contributed to the broadening of the concept of imitation through their reexamination of some key issues concerning the arts. The greater the Aristotelian influence, the more the emphasis shifted from the artistic object to the "contriving and considering of how it may come into being"[26] and to the "import" it harbors. Art, for Aristotle, was a kind of "potency" a productive form of knowledge, while imitation partook in the cognitive process of *analogy*, a process which he considered a part of all learning. By emphasizing the productive aspect of art and by circumscribing imitation, as he did, by "purpose or design," Aristotle himself helped turn "imitation" into a polysemic concept. Aristotle, thus, provided the arts not only with a powerful concept, but with one open enough to accommodate many subsequent queries.

Entire gamuts—from likeness to symbolic representation, from impression to expression, from conception to perception, from common traits to the uniqueness of the arts—could be entertained under the umbrella of "imitation" as conceived by Aristotle. As it turned out, he vested the concept with the kind of potency whose offsprings he himself could hardly envisage. In contradistinction to Plato, Aristotle opened the road to illusion and the *realities* which it conveys by legitimizing the fictive. It is no coincidence that Twining and Smith, last but not least among our writers, should have returned to the cognitive and aesthetic examination of the "nature" of imitation in "what is called the imitative arts." It is by no means an exaggeration to state that their understanding of the aesthetic constraints in the production and consumption of art created the *conditions* necessary for entertaining the conception of modern art, while their analysis of cognitive constraints contributed to the modern views of art *in general,* whether philosophical or psychological.

At different points in our discussion, we have alluded to a process of demystification. By way of summary, one may claim that the process as a whole encompassed, broadly speaking, a transformation of the relationship between the "given" and the "made," that is, between Nature and Culture. While terms like nature, imitation, expression and perfection continued to be employed, their meanings underwent significant changes, resulting from, as well as effecting, new qualifications within the twin domains of the given and the made. The rising interest in perception, and the reformulation of the epistemological standing of causal explanation, led to the employment of two distinct procedures, the one related to the reexamination of the "given" and the "made" as wholes, the other referring to their internal investigation. While the former strove to unveil common denominators for each of the domains, the latter tried to reveal differences that helped define uniqueness within them. Interestingly, the more one focused on the "made," the more differentiated it became, even if many of its aspects still awaited definition. In the course of such clarifications, much of the "natural" inadvertently underwent redefinition, relocating much that was previously considered as given. For example, many of the queries which found their solution in the *multiplicity* of the "naturally given faculties" were now reexamined in light of the *unity* of the mind, which called for better understanding. Paradoxically, such an understanding was expected to enhance the grasp of the "made," which depends on perception and cognition, at one and the same time.

The immediate consequence of this development is to be seen in the reduced discussion concerning Art in general, and increased attention to the uniqueness of each of the arts, and the ways they may be manipulated so as to transcend their individual boundaries. Indeed, artistic *literacy* and *informed* taste, now merited the kind of recognition which they had never enjoyed before, stressing less social credentials, in favor of "benefits" free for all who wish to be *enlightened.* Partaking in culture turned into an active role, for it became clear that "perfections" of all kinds are not only circumscribed, but also largely effected by cultural desiderata. Artistic expression was likewise seen to involve the partaking agent, for the artistic import embedded in artistic objects can be energized only in the affect to which it gives rise, which, in turn, resides in the consumer. While it is fair to admit that the arts may have lost some of their mythical powers in this process of intellectualizing, it is equally fair to claim that they emerged as *systems* that justify serious philosophical and scientific treatment. The rigorous treatment of art was essential, above all, to the understanding of the role of *illusion* in art. Dispelling beliefs, opened the road to an understanding of the *nature* of the "suspension of disbelief."

Notes

1. "Music can noble hints impart,/Engender fury, kindle love;/ With unsuspected eloquence can move,/ And Manage all the man with secret art," wrote Addison in his "Song for St. Cecilia's Day, 1764, at Oxford" (Addison 1914: 22), expressing the conviction in music's secret power over the passions. Winn (1981) tried to subsume under the title *Unsuspected Eloquence* the history of the relations between poetry and music in the West.
2. The exchange of techniques and procedures between the two arts is central to many studies, though not all deal with the subject directly. Of the immense literature, let us mention but a few relevant examples. For general studies of the subject see Winn (1981) and Stevens (1986). For Medieval and Renaissance treatises on the subject consult Harrán (1986). Spitzer (1963) discusses the influence of musical ideas on the development of central stylistic procedures in poetry, particularly the rhyme.
3. For the process that led to the musicalization of texts, which eventually served as libretti, see Katz 1986: 135-175.
4. By contrast, Winn (1981) views the issue in pejorative terms. He considers the English treatises as "humanist attitudes blurring into Romantic ones without the reexamination of musical logic that contemporary developments in music would seem to have demanded" (252). Heavily relying on Schueller's account, and particularly concentrating on John Brown's Dissertation of 1763 (Brown 1763), his criticism is not surprising. He totally missed out on the importance of Webb, and also ignored other writers whose innovativeness can hardly be doubted.
5. The term was coined by Nelson Goodman, and serves, as well, as a title of his book, *Ways of Worldmaking* (1978). The notion, however, with some modifications, is shared by many contemporary thinkers, philosophers and historians.
6. Kant published his third critique—in which aesthetic considerations occupy only a minor part—in 1790. By contrast, Twining's treatise, the latest of the treatises addressed in this volume, was published in 1789. Schelling's, Hegel's and Schopenhauer's writings appeared in the first decades of the nineteenth century.
7. According to Kristeller (1965), it was the *Querelle des Anciens et Modernes*, which brought to the fore the recognition of the virtually two distinct bodies of knowledge—science and art. The contenders of the *querelle*, argues Kristeller, led to the insight "that in certain fields, where everything depends on mathematical calculation and the accumulation of knowledge, the progress of the moderns over the ancients can be clearly demonstrated, whereas in certain other fields, which depend on individual talent and on the taste of the critic, the relative merits of the ancients and moderns cannot be so clearly established but may be subject to controversy" (193-4). Discussing the development which led to the crystallization of the new system, Kristeller claims that certain concepts and attitudes, which later proved to be of importance with regard to the new classification, were transmitted from the Classical Age. The Renaissance, he argues, "brought about many important changes in the social and cultural position of the various arts and thus prepared the ground for the later development of aesthetic theory." Yet, the Renaissance, as he himself observed, "did not formulate a system of the fine arts or a comprehensive theory of aesthetics" (178). Toward the end of the sixteenth century, more frequent alliances of the three major arts (i.e., poetry, painting and music) are in fact mentioned, yet, all through the

seventeenth century, Kristeller notes, the scheme fluctuated. Among other factors which led to its stabilization in the eighteenth century, Kristeller included the rise in prestige and appeal of painting and music, due to "an amateur public to which art performances were addressed" (225).

8. Interesting in this connection is Monboddo's treatise *On the Origin and Progress of Language* (1773-92) which brought some of Rousseau's ideas to the British sphere, supported by the then new anthropological findings. See Lovejoy 1933.
9. See Townsend 1982. Smith, of course, will show that art never frees itself from basic interests inasmuch as it is a protagonist in the free market, which affects aesthetic qualities.
10. Already Berkeley, somewhat earlier, understood that vision cannot be explained other than in a nondirect fashion. In fact, the latest theories of nondirect vision, like those of Arnheim (1969), Ullman (1980) and others, may be traced back to Berkeley.
11. Hutcheson 1973: 81; *AM*: 74. The Italics here and in the following quotations from the British treatises are in the original.
12. Willy (1961) traces this transition in illuminating brushstrokes.
13. Caygill (1989) brilliantly exposes the problems and incongruities of such a self-serving agenda.
14. "I may say of music," writes North in the first *Musical Grammarian,* "as the young man in the Gospell sayd of the Law, and (perhaps) with like arrogance, that all these things I have observed from my youth upwards. A musicall family, conversation with the best masters of the time, and a pittance of formall teaching, made me a professor; and from a medler with most sorts of instruments (not excluding the voice), a buisy body in transcribing, and composing, and a meeting-hunter of musick in London for divers years, I found my self drawne in to such a familiarity with the *prattique*, that performing at sight, was to me like speaking, walking, or running, and there lyes the secret of a true lover, who is never so undone, as when his mistress is easy, and smiles" (Wilson 1959: xxii).
15. "To the writings of his essays, North brought an inquiring and realistic mind, warmed always by a spontaneous love of music and a ready response to its appeal. Music was 'a splendid science' to be explored and explained in a new age of Reason, and it was also, for him and his circle, 'our dear art'. This double approach—the speculative and the practical—was characteristic of the *virtuosi* of the restoration period, and in North's case is to be found not only in his music, but also in his study and practice of drawing and building, in his inquiries into physical science, and even in less likely fields such as that of estate-management. Always with the practice went an immediate interest in the theory, and to find and state the 'reason' was for him a foremost aim" (Wilson 1959: xxiv).
16. See our discussion of Lessing's theory in chapter 6.
17. "Let us speak of work of art as *autographic*," Goodman suggests, "if and only if the distinction between original and forgery of it is significant; or better, if and only if even the most exact duplication of it does not thereby count as genuine... Thus painting is auotographic, music nonautographic, or *allographic.*" About the latter he says: "Performances may vary in correctness and quality and even in 'authenticity' of a more esoteric kind; but all correct performances are equally genuine instances of the work" (Goodman 1968: 113; italics in the original).
18. Jacob's "give and take" among the arts throws new light on the *ut pictura* tradition: "The *Painter* is to *animate* a *Form*, and the *Poet* to lend a *Form* to *Sentiment* and *Diction:* One is to give *Life* to beautiful *Proportion*, and the other *Strength* and

Figure to sublime *Thought*." Jacob, as expected, attributes this reciprocity to general aesthetic principles which all the arts realize: "As these *Arts* proceed chiefly from the same Principles, *Imitation* and *Harmony* so they are mutually assistant to each other, and ought to dwell much together; yet however they may be reciprocally oblig'd to each other, and agree so well in the main, they have their *separate* Beauties too" (*AM*: 85).

19. The Art Symbol, according to Langer, refers to the work as a whole, which does not stand for something else, nor does it refer to anything that exists apart from it. It formulates experience which discourse is unable to articulate. The work of art as a *whole* does not point to meaning beyond its own presence. What is expressed cannot be grasped apart from the *sensuous* form through which it is expressed (Langer 1957). While this conception can be criticized as organic romanticism dressed in modernist attire, it can be salvaged nonetheless under notions such as syntactic structures, archetypal constructions, thereby avoiding metaphysical overtones.
20. That Webb should grant high status to the critic is not surprising, given the estimation a critic like Warburton received, which was acknowledged to be higher than that of the poet (Lipking 1970: 340-2).
21. See Omond 1907. Among the treatises, which dealt with meter and number, many were concerned with the metricality of blank verse and even tried to supply the reader with an appropriate notation (e.g., Cooper 1745; Manwaring 1744; Mitford 1774).
22. Webb could have found a source for his ideas in Quintilian, who argued that "the ancients saw that pleasure expands in the soul, in the thymic part, pain in its descendants, and in the irrational part, divine suffusion" (Quintilian 1983: 121). In light of the above it is clear how Webb worked this basic metaphor into a sophisticated argument, which takes into account not only the inherent qualities of each field—music and the emotions—but the way by which they come to be cognitively associated as well.
23. This passage is taken from the second part of Dissertation I of Twining's treatise (Twining 1789: 35), which is not included in our selection. Italics are as in the original. The whole part is devoted to the ways in which the word imitation was applied by Aristotle; more particularly, to why Aristotle abstained from discussing *kinds* of imitation (e.g., descriptive imitation), which for Twining seemed so essential. Gombrich in his illuminating "the Renaissance theory of art and the rise of landscape" (Gombrich 1985: 107-121) supports this contention, giving ample historical evidence.
24. The autonomy of music bespeaks its direct expression: In contrast to dance, "music and poetry" argued Smith, "can each of them subsist alone," moreover, music consists of "unique combinations of sounds," it is unique in its possession of some kind of original and independent beauty, not to be found in reality (*AM*: 393). His argument is reminiscent of Kant, only that Smith ascribed to music more cognitive value.
25. See our discussion of metaphor in chapter 2.
26. Aristotle, *Nicomachean Ethics,* book VII.

6

The French and German Scene

Our case for connecting aesthetic theory to cognitive science rests, as we have seen, on the assumption that the eighteenth-century British school uniquely anticipated the modern view of art as self-referential, and through comparative study of the attributes of the several arts, foresaw the emblematic role of music as bearer of syntactic, rather than semantic, "messages." Before concluding therefore, we must address a question that hovers over our claim, namely, why did the British "do it" whereas the others did not.

We cannot allow ourselves to delve into the history of the rich development in music, literature, and literary criticism of the Enlightenment, nor even into its aesthetic theorizing. To do justice to the question involves no less than a complete survey, similar to this one, of the deliberations on art in the major cultural centers of the time. While we cannot, obviously, do this thoroughly, we may try to explain why and how the Italians, the French, and the Germans differed from the British.

Though the Italians had a very active musical life, exemplified both in their operatic productions as well as in all that pertains to the rise of the classical symphony, they were not occupied, comparatively speaking, with philosophical deliberations concerning art, in spite of their dedications, forwards, encomia and the like. Of course, they also voiced opinions about what they and others were doing. But important as were the contributions of people like Padre Martini, Algarotti, Marcello, Arteaga, Muratori, and many others, they did not confront the problem of the epistemological and cognitive standing of the arts, certainly not to the point of providing a significant corpus of a unique character.[1] The French, by contrast, did leave a mark of their own, in the philosophizing about music, but the real enigma is Germany. After all, it was German soil that nurtured the growing understanding of aesthetics as a bona fide branch of philosophy. Likewise, nobody can deny the central role of Germany in the newly acquired autonomy of instrumental music, which, of course, compounds the puzzle, since it is precisely the autonomy of music that was so central to the English deliberations concerning the aesthetics of music. Though the Germans lagged behind in matters concerning musical perception, they did contribute, in other ways, to the overall issues with which we are dealing. To what channels did they turn their creative thought? How did the intercourse with the English and the French affect their discourse? We shall try to cope with these questions in the following pages, hoping to throw some light on the European picture while enhancing the English scene with the perspective it needs. But let us consider the French first, then the Germans, always bearing in mind that we are trying to understand the thinking of each about the art of music as compared with the thinking of the British.

Scientific "Fundamentals"

The deliberations concerning music in France did not invoke the kind of comparisons among the arts, and attempts to understand and define the uniqueness of each that characterized their counterparts in England. With few exceptions, they grew primarily from discussions centering on Rameau's theoretical writings, which covered a wide range of issues concerning music, art, and science.[2] The debates over Rameau's writings on music engendered a discourse that addressed more fundamental philosophical issues, such as the nature of artistic expression. Such issues, however, were closely related to yet broader issues concerning the respective roles of reason and experience in both science and art, and eventually led to a reexamination of both universal and cultural principles.

In the Age of Enlightenment, France was the center of ideas, yet the thought of the period was not uniquely French; it drew from a variety of sources. Though skeptical toward tradition, it carried over the philosophy of natural law, which held that right depended on universal reason. From the intellectual revolution of the seventeenth century it carried over not only Descartes and Newton but Bacon and Locke as well. The empiricism of Bacon and Locke, as we have seen, had severely challenged the medieval deductive method of inquiry. Truth, they suggested, was an end inductively arrived at after investigation and experiment, rather than something that is postulated a priori. At the same time, the so-called "Cartesian dualism," that is, the two fundamental "realities" in the universe—the "thinking substance" (mind, subjective experience, consciousness, etc.) and the "extended substance" (everything that is "objective") was fully accepted. Moreover, everything except the mind itself rested on a universal fact, namely, that it occupied a portion of infinite space and could be calculated according to the geometry of Euclid. Whereas aspects of the mind seemed elusive, "extended substance" was quantitative, measurable, and reducible to formulas or equations. Mathematics, accordingly, was the most powerful instrument available to the human understanding.

The success in uniting disparate theories under one conceptual roof resulted, of course, from the basic endorsement of the Cartesian dualism. However, the application to intricate and complex systems of that which was created out of the pressure to withstand incommensurability often involved kinds of analogies which had no scientific standing whatsoever. For example, the use of the theory that "particles are attracted to each other" (related to the basic idea of universal gravitation)[3] in the framework of social thought was problematic, to say the least. Confusing the objective with the subjective, passive bodies with active agents, rendered the "regulating force" mysterious. Indeed, the simple application of scientific models to social and psychological domains often resulted in kinds of metaphors unacceptable in bona fide science.

The confusion between the objective and the subjective, between nature and culture, applies also to Rameau. It is not surprising, however, that Rameau should have been considered, nonetheless, as the "Newton of Music." He was indeed able to attach himself to the world of science owing to the physical nature of music. Moreover, he was in a position to integrate disparate musical theories within an overall deductive theory that aspired to a high degree of lawfulness. However, ignoring both the cultural settings that give rise to different kinds of music, as well as the psychological factors that affect its reception, he failed to treat music as art. The fact that Rameau believed that the physical and the cultural could be explained by the same scientific paradigm exposed his lack of understanding of the paradigm itself.

Rameau's attempt to superimpose the physical on the perceptual, to say nothing of the expressive, did not escape the criticism of some of the French men of letters—the *philosophes*. Descartes's doctrine of innate ideas—the inevitable disposition of the human mind to think in certain ways—had already been challenged, as we have noted, by Locke's theory of knowledge. Locke maintained that knowledge is derived from *experience*, from the perceptions of the sense organs and the reflection of the mind on these perceptions. The *philosophes* were, of course, aware of these ideas. While contributing to the popularization of scientific ideas, they themselves struggled with them, attempting to reconcile conflicting ideas for the sake of a bourgeois middle class that was becoming educated. They were concerned not only with purely philosophical issues, but also with political and social issues, which they honestly tried to address. But here, too, they were attuned to their neighbors across the channel.

In order to highlight the social and political context, in which the deliberations took place, it should be borne in mind that the eighteenth century was an age of spectacular enrichment and commercial expansion. In the global economy of the century the French held commercial primacy in the European markets, while the British virtually monopolized Europe's trade overseas. For France it was a period of absolutism checked and balanced by organized privileged groups, whereas in England the Parliament acted as the representative legislative body to which different kinds of groups, who were willing to go through its channels, could appeal. Neither system, of course, was free of corruption and injustice, but the privileged groups in France were politically less competent than their English counterparts, and the British parliament, moreover, proved an effective machine for the conduct of public affairs and business.

Whereas the notion of the public good is invariably echoed in the treatises of the English men of letters, it is Liberty, the liberty of the individual, and especially the freedom of thought, which guided most of their French counterparts. If Voltaire admired England, it was for its religious liberty, its relative freedom of the press, and the high regard paid to men of letters like himself. Once a government was enlightened, he did not care how powerful it was. Unlike

Montesquieu, who believed in the separation of the functions of the executive, legislative, and judiciary, and therefore admired the English constitution as he understood it, Voltaire had little objection to "enlightened despotism." Among those concerned with the problem of liberty, Rousseau was quite unique, for he was willing to challenge the premise that all men subsisted under the same natural law of right and reason. In fact, he questioned the belief in the potential similarity of all human beings. Though he, too, considered the rightly ordered state the best guarantor of social welfare, not believing in the unity of mankind, he sought to emancipate the *collective* will from social pressure, advancing the doctrine of the sovereignty of the people. Reason was not enough to secure a desired civilization, he maintained, the best traits of human character—kindness and true understanding—had to be taken into account as well. The presence of feelings, which are the most genuine product of nature, claimed Rousseau, had to be recognized for what they are. Sympathy, intuition, desires, not only critical thought, had to be trusted in a state in which every person could feel that he belonged. Man requires a sense of both membership and participation in public affairs, that is, in the common will.

As the "man of feeling," Rousseau became not only an important forerunner of romanticism, as is well known, but more significantly he became an important source of all modern emphasis on the nonrational. No wonder that Rousseau's contributions to the most serious enterprise of the *philosophes*—the *Encyclopedie*, the "reasoned dictionary" edited by Diderot and d'Alembert—encountered so much resistance. Thus, Rousseau's debate with Rameau, concerning the expressive means and content of music should not be viewed only as an attempt to shift the focus from harmony, as a musical element, to melody, which he relates to language. Nor should it be viewed as yet another attack on French music.[4] It should be understood, rather, as an overall attack on Descartes's mechanical universe, which conceived universal laws as known only through innate reason. If he championed melody, it is because he was concerned with a more direct form of expression that would convey underlying human emotions. Discussing music the way he did, betrays Rousseau's overall concern with the "unlogicized," trying to make room for "languages" of passion, circumscribing their functions in the life of human beings and the cultures which they create (Rousseau 1753/1966).

The theory of mimesis, as we have seen, though it was closely related to a conception of imitating human passions,[5] placed great emphasis, in the course of time, on the representational aspect of art, thereby beginning a search for a more specified understanding of that which is being represented and how representation is accomplished. Meaning, as we have noted, has to be acquired, and the acquisition of meaning, in turn, calls for an understanding of the modes of symbolization that bridge between representations and the meanings they convey. It is no wonder, therefore, that for many centuries imitation in music was far more problematic than it was in the other arts and invariably referred to music with text. Rousseau's attempt to deal with music as a *direct* and *immediate* expression, as a "natural language" of the passions, challenged not only the representational aspect of the musical art, but even the core of the theory of imitation, though he did not intend to overthrow it.

One may still entertain a theory of the universality of the passions, as Dubos, in fact, did, even if passions are no longer what one imitates, but rather that which one expresses. For Dubos, that which one expresses is itself anchored in a given quality possessed by expressivity; art, thus, reflects, nonetheless, a reality of sorts.[6] More inclined to identify art with *ideal* reality, Batteaux, too, viewed its *signifiers* as given rather than as created; on the basis of his *own* premises, Batteaux could indeed argue, employing Rameau's basic notions, that the signifying sounds and their derivative harmonic relations are of natural standing.[7] Despite the differences between Dubos and Batteaux, both had difficulties discussing music in terms other than those of the descriptive aspects of circumscribed situations. The expressive qualities of "absolute" music—Fontenelle's unanswered query—was still not served thereby.[8] By contrast, Rousseau allowed music to come into its own, emphasizing its a-rational, prelingual aspects. Though he did not wish to contradict Rameau's theory concerning the fundamental bass and its relationship to the structure and behavior of chords and chord progression in the tonal practice of the time, he nonetheless reduced it to a specific musical grammar, rather than viewing it, as Rameau did, as the natural basis of all music and musical expression.[9]

Rameau tried to establish that harmony as well as melody are given to man directly in nature. More specifically, what is given in nature, claimed Rameau, is the resonating body, which generates both the fundamental sound as well as a series of harmonically related overtones, which constitute the consonant intervals of harmony. True, harmonic relations are experienced instinctively, but they are still a part of the mechanical rational universe which can be explained through mathematical calculations; rather, the latter explain their immediacy whereas the effect which we experience constitutes a reference for whatever changes may be introduced to its initial cause. Instinct and reason are not opposed to each other, but confirm each other as two sides of the natural basis of expression in music. Moreover, since the source of expression comes from nature, the effects are universal.[10]

Rousseau's interest in the languages of passion was less concerned with universal qualities than with human nature and cultural differences. What remains universal however for Rousseau, the human need for self-expression. In his hypothetical reconstruction of the beginnings of language and music, which appears in his *Essai sur l'origine des langues où il est parlé de la mélodie et de l'imiatation musicale*(1753), Rousseau argued for the common origin in a presocietal state of nature. The first "language," Rousseau suggested, was expressive, rather than discursive, a sort of union between speech and song. Melody, in fact, still "speaks" this language and is able, therefore, to communicate passions rather than reason. And, since melody is natural for *human* expression, its meanings, unlike those of the physical world, invariably depend on culture and society (Rousseau 1966). If the Italians excelled in their music, writes Rousseau in his *Lettre sur la musique françoise* (Rousseau 1753), it is because of their affinity with the conditions in which the passions find vent in song. This is why Italian is a more musical language than French, and lends itself more readily to musical renditions. Like some counterparts in other countries, he too arrived at the conclusion, albeit in his own unique way, that the expressive powers of music surpass those of the other arts because of music's "directness," stemming from its unfettered vocal nature.

Rousseau was undoubtedly influenced by Condillac, and it is interesting to compare the two.[11] Condillac's main thrust, as we have seen, was directed to rectifying Locke's theory of knowledge, or more particularly, its implications for the developmental aspects of linguistic acquisition. His interest in music as "a language of action" was thus related to music's *gestural* abilities which he conceived as prerequisite to the synthetic aspect of language, that is, linguistic utterances. The latter complemented his other contentions with regard to the specific lexical functions of words. Though the language of action, according to Condillac, is at all times a necessary component in the cultural functions of language, it preceded all other elements of language in the stages of its development. Whereas Condillac made room for music even in his social and cultural theories, he did not accord an independent standing to music's own developmental stages—those which enabled it to achieve the "perfection" that Jacob, for example, spoke about. Interested in music as such, despite his preferences for one or another type, Rousseau was willing to grant this perfection to music of all times, even if he did not make clear in what it consisted. In his search for transparency in human relations and authenticity in their personal expression, music was accorded a central place (Starobinsky 1988).

The elusiveness of Rousseau's explanations and Condillac's speculations could not easily be sanctioned by people who were more scientifically minded. D'Alembert, for example, though he denounced Cartesian metaphysics and all kinds of underlying causes, was willing to trust only that which could be established, beyond doubt, through careful observation and experimental rigor. It is no wonder that he was ready, at first, to support a theory which focused on music as a physical phenomenon (setting aside music as *art*), rather than a theory based on hypothetical conjecture. D'Alembert too, however, moved further apart from Rameau, as Rameau moved toward converting his original theory into a more universal system. Upon Rameau's eventual insistence that his "proportions and progressions" are the basis of all the arts and sciences, d'Alembert not only withdrew his support, even concerning its usefulness for music, but reopened the question of the very nature of music theory.[12] There is no basis, he claimed, for expecting conclusive evidence for *absolute* principles, not only for music as art, but even for music as science (Verba 1993: 63-66). In this sense, he shared Rousseau's early recognition of the limitation of any positivistic approach to music.

Diderot made an attempt to reconcile the objective and the subjective approaches to the artistic experience.[13] As chief editor of the *Encyclopedie*, he experienced the pressure to maintain some consistency in his enormous undertaking. Aware of the increasing role attached to sensibility in artistic expression and to empiricism in scientific investigation, he, too, eventually rejected Rameau's theory. He was aware of the complexity of artistic experience but was unable to disentangle it. He, too, stressed the differences among the arts and the special challenge that music presented to the theory of imitation, but did not carry this understanding of the arts far enough. His own attempt to construct a more systematic aesthetic theory, consequently, failed. Yet, he was aware of the tension between general laws and individual responses, and was also willing to entertain the consumer's perspective.

None of these men of letters, including Diderot, approached aesthetic problems the way the English did; some even began to doubt whether a rational justification of aesthetic judgment was altogether possible. Boyé denied music's semantic ability because of the failure of positivistic theories to explain its expressive powers in rational terms.[14] Chabanon's dissatisfaction with the unsuccessful attempts on the part of his countrymen to provide better theories, led to his emphasis on the sensational constituents of music, and to his denial of their cognitive standing.[15]

Altogether, though the major philosophical debates about music in eighteenth-century France seem to have been based upon enlightened premises, some ghosts from the past were still commanding the scene. In a rather ironical way, Rameau attempted to unify Descartes's deductive approach with Mersenne's empirical findings.[16] He was able to do this, assisted by the Classicistic belief that artistic norms were equivalent to scientific axioms. His position in the Royal Academy of Science symbolized and reenforced this peculiar mix of theoretical notions. While his colleagues in the academy began to be critical, sensing the problematic it entailed, they could not entertain Rousseau's views as a real alternative since his views were still cast after the mystical *je ne sais quoi* approach.

We should like, nonetheless, to conclude our discussion of French musical thought with two examples that did involve comparisons among the arts in an effort to clarify the uniqueness of each. The first is Rousseau's somewhat haphazard comparison of music and painting, the second is Chastelleux's comparison of poetry and music.[17] Rousseau tried to clarify the difference between *mere* sensations and perceptions that possess cognitive content. Identifying the latter, not unlike Condillac, with images and signs, he denies colors a cognitive role; representation, he implies, functions as a unifying factor, which provides a picture with its artistic standing. Music, too, requires such a unifying agent, he argues, thus equating melody with figure, and harmony with color. While risking thereby the reintroduction of "mimesis" into the discussion of music, Rousseau unwittingly admits a sort of "sense formation without predication," making room for such formations not only in music. "Correspondences" (among the arts) notwithstanding, Rousseau is careful to introduce "limits" (of each individual art) into his discussion. Aware of the spatial nature of painting and the temporal nature of music, he points to the illusive presence of the human factor in music, related not only to the production of sound but to its reception as well. He is also in a position to understand the independent status of colors as compared with the relative status of musical sounds. Rousseau is willing to endorse the transgression of the limits of the artistic medium, at least as far as music is concerned, the very point Lessing will dispute in his *Laocoon* a decade later. Music, Rousseau tells us, can represent the "imperceptible" aspects of the visible: "The musician's art," he writes, "consists of substituting for an imperceptible image of the object the movements which its presence excites in the heart of the contemplator" (Rousseau 1966: 64). If artistic imitation, as we have learned from Baumgarten, is mainly related to the effects of reality on the mind, rather than to reality itself, music is able to imitate the effects of these effects. The French tradition of musical "painting," in practice as well as in thought, presents here an interesting turning point.

These ideas of Rousseau are close to Smith's overall aesthetic conception; indeed, Smith's discussion of musical imitation includes a long quotation from the *Essai sur l'origine des langues*. Yet Smith also tells us, in no ambiguous terms, what he thinks of the famous French writer: Rousseau, he writes, is "more capable of feeling strongly than of analyzing correctly" (*AM*: 410). Had Smith been more explicit, he might have pointed out that Rousseau failed to account for the special symbolic nature of the musical data, for which Webb, as we have seen, had a fine explanation. The fact that Rousseau's position regarding mimesis remained obscure, erupting again in the famous debate between Derrida (1967/1977) and Paul de Man (1983), stems from this basic deficiency.

Chastellux's essay concerning the union of poetry and music, important as it may have been at the time, did not provide new insights as far as aesthetic theory is concerned. The essay constituted a part of the ongoing debate concerning the relative merit of the Italian vs. the French approaches to opera. Though Chastellux was committed neither to the one or the other, he was willing to admit that the Italians more often "avoid the impossible," that is, they try not to wed that which cannot be wedded. Since music has its own internal dictates (repetitions, elaborations, all kinds of formal requirements) the union between poetry and music, says Chastellux, invariably affects the *choice* of the text. That does not mean that the text becomes irrelevant; on the contrary, it highlights the function and contribution of each medium to the success of the union between them. The productions of the lyric stage, he tells us, require not only the close collaboration of musician, librettist, stage designer, etc. but their mutual understanding of each other's role in the bringing about of a unified work that will do justice to its dramatic development. Chastellux does not discard ballets, little airs, or stage machinery of sorts, he asks only for a more *organic* integration of these, in order to avoid mere displays. This also explains his insistence on "proper" places for arias, duets, etc. since their misplacement may suspend the logical development of the action, which is so essential in the unfolding of the whole.

In spite of Chastellux's many valuable suggestions regarding meter, declamations and the like, there is not much novelty in what he has to say about either poetry or music. He does, however, try to circumscribe the "limits" of the individual arts and suggests ways in which they ought to accommodate each other when united. This extended essay on the subject, full of examples, constitutes, nonetheless, a continuation of what had already become an established tradition, namely, to attack, defend or try to understand opera as a *medium*. Yet, such attempts to understand the uniqueness of opera invariably cast music in the role of the carrier of sentiments, however ill defined.[18]

What we have seen so far reinforces the claim that aesthetic judgment called for a new kind of reasoning based on a totally new orientation. In retrospect it seems that the more pragmatic and down-to-earth approach of our English writers, akin to the inductive method preached by their fellow countrymen, enabled the breakthrough we are discussing. Even if their writings contained some speculative thought, their readiness to deal with the senses and sense perception concerning the arts opened a new avenue towards a better understanding of the aesthetic domain.

Religious "Overtones"

Even provisional answers to the question we posed take us into cultural contexts and historical processes wider than the question seems to allow. Thus, discussing the German case, one has to take into account that the "Germans," at the time, unlike their neighbors in the west and the south, lacked the rudimentary basis that gives rise to a shared culture. The bifurcation of the "German soil," among Prussia, the Habsburg Empire and the manifold small German principalities, constituted the reason for this state of affairs.[19] Nonetheless, from imitations of foreign models, at the beginning of the century, the Germans arrived, towards its end, at a creative achievement of European significance. Not that borrowed models were completely discarded; on the contrary, the overwhelming growth of German culture, particularly in the fine arts and aesthetics, resulted, to a large extent, from the attempt to integrate different influences into shared legacies and latent trends. The routes traversed from Wolff and Baumgarten to Kant and Fichte in philosophy, from Opitz to Goethe and Schiller in literature and literary thought, and, somewhat differently, from Bach and Mattheson to Mozart, Haydn, and Beethoven attest to this development.[20]

Indeed, integration, rather than simple importation or original creation seems to distinguish the German "code" for a long time.[21] Without identifying the political, social, and economic antecedents and expressions which might have "constituted" this particular behavioral pattern we wish to circumscribe some of the cultural trends and undercurrents which affected the integration itself, determining its nature and character. With origins in the sixteenth and seventeenth centuries, four interrelated factors seem to have had a bearing on its development: (1) the pluralistic bent of Protestantism; (2) the influence exerted by Lutheranism on musical thought and practice; (3) the legacies of hermeneutic practice concerning the Holy Scriptures[22]; and (4) the tendency to adhere to inherited traditions and old beliefs while accommodating new ideas.

Unlike the English and some of the French, the value of music *qua* music was never doubted by the Germans. Paradoxically, Luther, who "lowered" music to a declamatory function, never doubted its self-sufficiency. The enjoyment experienced while listening to music, he argued, is neither evil nor accidental (as some Catholic authorities believed it was) but "a gift from God."[23] Moreover, Luther and his adherents believed that "lifting the soul to God" could benefit from music even without the assistance of words. Luther's contentions, that "none of the other art forms can hold a candle to music" and that words are incapable of "singing" music's praise, were germane to the German myth of music's ability to express the ineffable, a myth which has withstood the changes of times. Whereas Luther equated music with theology, two and a half centuries later, the Germans, enhancing the myth, substituted religion by musical reverence and devotion, Dahlhaus tells us (Dahlhaus 1989: 74-7). One should not overlook, however, the importance Luther attached to musical education: "It is vital and necessary to keep music in the schools," he wrote, "a schoolmaster needs to be able to sing, otherwise I deem him a good-for-nothing." The latter had important repercussions for the musical culture that was eventually cultivated in Germany.[24]

The Lutheran "adoration of music" was, from the beginning, intertwined with the pluralistic character of Protestantism, which had its roots in the secession from the Catholic Church of various splinter groups, who wished to revolutionize religion as well as religious practice (Katz 1985). With similar objectives in mind, Luther sponsored the creation of a new canonic body of "songs"–the simple homophonic chorales, encouraging as well instrumental music of an exuberant kind. These two contrasting models, the one in which text dominates the music and the other, which altogether dispenses with text, existed harmoniously side by side in the Lutheran tradition.[25] Debates like those of Zarlino-Galilei or Artusi-Monteverdi are unthinkable in a tradition which never considered harmoniousness the antithesis of affectivity. *"Seconde prattice"*—brought from Venice, Florence or Rome—were always welcome as additions to the pluralistic Praise of God and Music. "*Stimmung*" expresses it all, as Spitzer has taught us, for it unites the general with the specific, harmoniousness with mood, "the stable tunedness of the soul" with a "temporary condition." It expresses "the unity of feelings experienced by man face to face with his environment (a landscape, nature, one's fellow man) comprehending and welding together the objective and the subjective into one harmonious unity" (Spitzer 1963: 5).

This basic term and its derivatives (*Einstimmung, Gemütsstimmung, zusammenstimmen*) are uniquely German and reflect, perhaps, their deep conviction with regard to a "preestablished" connection between nature and culture, God and man, which is so profusely elaborated in their writings. While endorsing the new science, the Germans were hardly occupied with the kind of iconoclasm, which characterized other countries. Old legacies and beliefs continued to thrive in Germany not only in universities and schools (as we have pointed out in chapter 3, in connection with rhetoric) but also in philosophical treatises still attached to scholastic formalism. Nor did the Germans discard old cosmographical models; they were bent, instead, on improving them, taking note of new methods and discoveries, reshaping their world of knowledge rather than replacing it. Kepler's unique integration of Pythagoreanism, theology, modern mathematics and new methods of observation and measurement is a case in point. Kircher's encyclopedic writings, more bizarre and of lesser importance, exhibit a similar pattern.[26] Leibniz, too, trying to bridge logic and ontology, the unlimitedness of thought and the limits of the given, encompassed it all in his "possible worlds," allowing for different kinds of images of knowledge.[27] For these men and others, Pythagoreansim, however embedded in Platonism or neo-Platonism, constituted a latent guiding assumption; it functioned in similar ways for most German musicians and writers on music for a very long time, even into the twentieth century.

The preoccupation of German musicians with rhetoric in general and the *Figurenlehre* in particular (discussed in chapter 1), can now be re-viewed in this perspective. From Burmeister to Mattheson, Scheibe, and Forkel, German writers never doubted that composer and listener are connected to each other through some "immediate recognition of the soul" of the musical message it receives. Whether they equated music with language or "beyond," they always trusted its communicative power. They tried to analyze music's constituents, so as to glorify that which they had already sanctioned, and to better educate those who were entrusted with that legacy, who were also expected to contribute to the opening of new channels for the uninitiated. In other words, they were engaged in a kind of hermeneutics—in providing detailed commentaries on what was already "canonized."[28] The fact that in retrospect their attempts can be considered as a contribution to the understanding of the inner logic of the musical work, of the concept of style and of the difference between music and language, should not obscure their more "applied" intentions and their underlying assumptions.

The leaders of the German enlightenment tried, from the start, to invest their own traditional values with the worldly beauty and enjoyment associated with foreign models (Sheldon 1975). This dual emphasis was important even for Gottsched, who tried to superimpose "Gallant modishness" on "Teutonic solidity" (Sheldon 1975; Blackall 1959: 160-1; 197-200). It is hard to say wherein that solidity consisted of at the time. Evidently there was a will to replace some of the German middle-class "stiffness," including their pedantic scholarship, with a more pragmatic urbanity and a "semblance" of social sophistication. Yet more than in other countries, the stylistic changes, and the debates which accompanied them, reveal the preoccupation of the Germans with artistic norms as part of a culture-making process. Philosophers and critics were active in this process. More often than not, they were the ones who actually formalized norms, or at least granted them the validity they needed for their cultural acceptance. To be sure, the association of artistic "norm" and "form" (to paraphrase Gombrich[29]) is immanent to art, yet crucial differences may appear in its consideration. German normativism is, of course, related to the French neo-classicism, which was normative throughout. Yet it seems that the French were more inclined to locate norms in a general conception of nature and human commitment, and less in values unique to cultural or artistic desiderata. Though Gottsched—the literary "lawgiver" from Leipzig—condemned all that was not made in accordance with the sanctioned French models, and though his influence was strong enough to render old German literature obsolete, the emerging German culture of the century was much more inclined to follow Baumgarten, Winckelmann, and Herder, who tried to mold new and different kinds of local norms.

Thus, Baumgarten's application of Leibniz's epistemology to the aesthetic realm meant that perfectibility in art is not related solely to craftsmanship and form, but is primarily related to the processes of thought which generate works of art and "reconstitute" them in the perceiver's mind. Neither models nor manuals constitute the basis of these processes, for they are mostly related to the perceptual domain and its dictates. The aspiration to perfection is no less essential for the aesthetic domain than for the "higher cognitive faculty." This argument, we claimed, was crucial for the eventual location of art in the mind. Yet the fact that it was logical thinking that served Baumgarten as an ideal for the epistemology of art undermined the very objective that seemed to guide him, that is, granting aesthetics the autonomy given to the logical and the scientific realms. Thus Baumgarten's impressive deduction of aesthetic rules should not obscure the fact that his statements took for granted that which needed proof and demonstration, that is, the conditions for the intelligibility of possible art worlds.

Baumgarten's German followers and critics may be viewed in these terms; they were busy constructing their theories of knowledge according to given axioms and postulates. Thus, Moses Mendelssohn, well versed in both French and English literature on art, insisted that the fine arts (painting, sculpture, music, dance, architecture, as well as poetry and eloquence) be reduced to a common principle, albeit better than "imitation."[30] Attacking Baumgarten for his concentration on poetry (a criticism later launched also against Lessing, and restated by Herder), he sought a comprehensive theory of art in the realms of goodness, truth and beauty, in line with the doctrine of the three faculties of the mind. Influenced by the Scottish philosophers, Mendelssohn's "faculties," though perhaps "higher" in status than the Hutchesonian "senses," suffer from that which the Scottish philosophers succeeded in circumventing by the perceptuality that they bestowed upon the aesthetic.[31] By granting beauty, and later also the sublime, a lofty standing, Mendelssohn preserved the work of art in its perfect form, keeping it from breaking up into its sensual substrata.

One should bear in mind that Mendelssohn was not only reacting to Baumgarten, Leibniz and Wolff whom he deeply admired; he in fact echoed Winckelmann's artistic ideals.[32] Whereas Baumgarten tried not to commit himself to *certain* forms and norms, concerned as he was with all kinds of possible artistic worlds, Winckelmann, by contrast, preached adherence to one ideal artistic world, associated with Antiquity. Obsessed with the latest discoveries of classical art, he seems to have created a myth that transcended the aesthetic realm. Indeed, his ideal of "noble simplicity and tranquil grandeur" ("*eine edle Einfalt und eine stille Größe*"[33]) formulated in connection with the Laocoon, encompassed moral and spiritual qualities as well. Winckelmann, in fact, advocated a kind of essentialism, stripped of adventitious ornaments and accidentals, of transitory passions and emotions and of superfluous movement (Nisbet 1985: 3-7). His overall ideals seem to have penetrated all cultural areas, attempting to retain the unity among the moral, the intellectual and the aesthetic spheres. Though related to different faculties, these three spheres were thus doomed to be interrelated even with Kant, who tried his best to separate their functions.

Even Lessing's anticipation of media theory, which attempts to relate the limits of artistic media to their inherent dictates, was still bound by a strict classical normativism. Taking issue with Winckelmann on the interpretation of Laocoon's posture and gesture, he nevertheless held the classical ideal of beauty to be the highest value of aesthetic judgment. Moreover, conscious as he was of "dimensions" in art, that is, of the "categorical" boundaries of time (its linear unfolding) and space (its simultaneous, gestaltistic presence) he could not entertain the thought of metaphorical exchanges and interactions among artistic domains. Each domain, he maintained, should adhere to its own dimensions as an epistemological dictate, determining the choice of artistic subject matter (See chapter 16 in the *Laocoon)*.

Compared with Smith's conception of domains and their domination, which could easily account not only for modern art but even for media not yet born, Lessing's essay seems strict and conservative.[34] No wonder that his views regarding music were likewise backward looking, though, on the whole, he did consider the indeterminacy of instrumental music as pleasing.[35] His conception of illusion is similarly "unimaginative," advocating a one-to-one correspondence with reality.[36] While it had important repercussions on nineteenth-century realism, the distinction he created between natural and arbitrary signs totally ignores the intricate ways in which perception turns into image, as Harris, for example, had already pointed out. Of the British thinkers only Twining—who was perhaps Lessing's sole counterpart as far as knowledge of classical sources is concerned—was on the verge of adopting simple realism. Discussing dramatic imitation, he almost seems to quote Lessing, with the significant difference that he was primarily concerned with the clarification of concepts rather than of norms.

At a time when the English had already forgone the classical ideals, having opted for a more relativistic cultural attitude,[37] the Germans were still preoccupied with what they believed to represent the epitome of humanistic achievements. Its distinctive features were constituted from the ethos of Hellenism merged with a middle-class morality, which had become fascinated with Rousseau and the myth of the "noble savage."[38] As an alternative to social inequality and political absolutism—the then current evils—this trend became, in due course, associated with the ideal of individual freedom that, as has already been observed, was central to the ideas of the *philosophes*. The German emphasis was different however: opting for *Bildung* and self-cultivation, they attached "religious" overtones of a devotional kind to the concept,[39] which, despite protests from Goethe and Schiller would eventually be merged with romantic ideals.

Indeed, the expressivism preached by Hamman and later by Herder weakened the strict hold of Hellenism. Yet their aesthetic ideology was still related to the molding of cultural norms, which found expression in the *Sturm und Drang* movement. Hamman's religious tracts, for example, offered the Holy Bible and human history as an imagistic, subjective alternative style, more closely related to the enigmatic and irrational aspects of the human soul. Poetry, all poetry, comprised for him a sacred phenomenon, "a natural kind of prophesying" which cannot be reduced to rules of sorts; its counterpart is the genius, who "plumbs all things, even the profound things of God" (Wellek 1981: 179-180). Attracted to the British philosophers (in a rather unphilosophical manner), Hamman employed their arguments, nonetheless, for the defense of his own feelings, intuitions and Christian beliefs. He emphasized the creative response of the reader as a key to the meaning of the text; whether or not he paved thereby the way to modern hermeneutics does not alter the fact that he failed to advance an understanding of *shared* behavior vis-à-vis works of art.[40]

Herder, his disciple, tried to reintegrate his master's ideas with the Hellenic ideals of beauty as promoted by Winckelmann and Lessing. His eclecticism notwithstanding, he advanced a new conception of art, which was supposed to replace the old mimetic theory, maintaining that art products should invariably be gauged against their cultural background, for they are expressive monuments of the community to which they belong (Nisbet 1985: 15-20). What might seem as a mix of Rousseau and Vico, was, in fact, yet another attempt to provide the Germans with a cultural orientation: He viewed *Volkspoesie* as the utmost expression of humanism, and sought to replace the French models, which he despised, by German folk models, especially those of Northern origin.[41] The new poetic genres which he proposed, together with new poetic ideals (Shakespearian tragedy, for example) and the new aesthetic concepts which he tried to develop—creative imagination, authenticity, organicism—exerted a deep influence on the ensuing Romantic movement.

Still, the changed agenda did not call for the discarding of sanctified norms. What it incited was, perhaps, a kind of typical German indulgence in correlations, interactions and synthesis of opposing concepts, subsumed by a unified outlook. Kant's inclusion of "the sublime," alongside "the beautiful," within his aesthetic theory, was perhaps paradigmatic in this regard; by relating each to a unique interplay between imagination and understanding, intuition and thought, the sensual and the abstract, he constituted a mode for dealing with two concepts, which are quite distinct from each other, within a larger framework.[42] Of course, if two such dissimilar ideals are reconcilable, why should the field not be opened to other divergent ideas as well, provided that they may be subsumed by the same meta-aesthetic principles. The aesthetic of the "naive" and the "sentimental," related to the artist's basic existential attitude (heralded by Schiller in his famous treatise[43]) added new categories to the strict dichotomy of "ancient" and "modern." Though Schiller basically identified the "naive" with Hellenism and the "sentimental" with modernism, his categories could also function independently.[44] Considering Goethe a naive poet—intuitive, spontaneous, and sensuous and himself a sentimental one—reflective and deliberative, Schiller betrayed his more immediate cultural concern: Schiller considered art as a basis for reconciling various dichotomous concepts, which the "harmoniousness of man and culture" requires. Unlike Rousseau, but like most of the German writers that followed him, he was always forward looking. Aware of the precarious situation of the modern individual and unable to forgo the typical German normative attitude, he considered all cultural stages, including the present one, necessary to the achievement of a desired perfection.

The Kantian "free" vs. "dependent" beauty is a more systematic example of a dichotomy, less culturally laden. His ranking of these two kinds of beauty is, nonetheless, typical. Whereas "free beauty" is Kant's own version of the basic Hutchesonian definition, "dependent beauty" introduces cognitive content, which is lacking in the objects of the first category. Kant needed artistic concepts as "food for thought," for he was trying to correlate the logical and the aesthetic. Like Baumgarten, he left the underlying process of aesthetic creation and aesthetic judgement largely indeterminate. Granting "imagination" a central place in his basic theory of knowledge, Kant argued that the artistic play of the imagination, realized in an ideal relationship between object and concept, is free by definition. Important as these arguments were for the epistemology of art, it again left the "explicandum"—the processes related to the creative imagination—virtually unexplained. Art works, according to Kant, are the products of "genius checked by reason," yet his conception of Genius, like that of previous thinkers, attributed to the new cultural hero subconscious creative abilities that can neither be objectivized nor analyzed.

On the whole, dichotomies make room for stylistic change and for the assessment of their cultural function. Nineteenth-century dichotomies and the dialectics they harbored, like classical vs. romantic, phenomenon vs. noumenon, Apollonian vs. Dionysian, in fact added new sub-concepts to the older ones so as to further facilitate the relocation of art in a changing reality. Still, handling artistic growth with an ever-increasing conceptual refinement did not significantly contribute to the clarification of its cognitive content.

Against this general background, we may try to assess, in broad terms, the developments in music and musical thought in eighteenth-century Germany, for the issues we raised have a direct bearing on these developments. Reviewing the contribution of people like Mattheson and Scheibe, considering the immense influence of French Gallantism, their emphasis on melodic "naturalness" and their attack on "turgid" styles (e.g., Bach's style) loom large.[45] Yet, even these people, to say nothing about those whom they attacked, showed clear German preferences. First of all, their outright reverence for music is never questioned and ever present. Mattheson, for example, went to great length to associate (etymologically) "*Musik*," through the Greek "muse," with the Hebrew word ???? (Maa'sse=act, deed), which in its biblical context is associated with God's creation (Mattheson 1739/1981: [Part I, chapter 2 §4] 85). At any rate, neither Mattheson nor Scheibe nor, for that matter, almost any of the German enlightened music theorists, held instrumental music (excluding virtuosic music), inferior to vocal music. If they saw fit to reduce the traditional *Figurenlehre* to but a few figures and tropes, it is because they considered the system either cumbersome or self-evident with regard to a style that had already become rhetorized through and through.[46]

As far as music making is concerned, it, too, resulted time and again in a unique integration. If Handel is considered as "the co-ordinator of national styles"—to paraphrase Bukofzer (1948)—Bach stands as their great "fuser." Facing the limitation of the gallant style, the Viennese composers, later in the century, sought in counterpoint—always considered part of their basic training (Kirkendale 1979; Webster 1974)—a better musical cohesion, without forgoing, however, the articulative power of the current style. While fuges and fugati gradually found their way to their symphonies and string quartets in the crystallization of their mature style, it was Bach's *own* integrated style, which served as an important model. Indeed, the finest works of the High Classical Viennese are inconceivable without the conscious attempts of composers to merge the "most German of all musical forms,"[47] that is, the fugue and its texture, with the cantabile, the musical articulation imported from Naples, Milan, and Venice.

The theoretical support for these musical developments, however, came from the North. In Berlin, where an "enlightened school" flourished, serious attempts were made to apply Baumgarten's aesthetic theory to the "art of sounds," granting specific musical meaning to his basic concepts.[48] Of course, the application of the theory to music derives from the growing recognition, initiated by Baumgarten himself, that "all of the arts and sciences have certain rules in common... yet each of them has its own peculiar percepts and beauties."[49] Likewise, Baumgarten's ideas with regard to the "confused" nature of art were conducive to a view that idealized its ineffable value. More specifically, Krause, for example, applied Baumgarten's dichotomy of "extensive" (perceptual) vs. "intensive" (logical) clarity (see p. 173 above) to the melodic vs. the contrapuntal dimensions, in support of the textural preference of the classical style. Unfortunately, the need to justify the analogy itself in perceptual terms did not occur to Krause.

Indeed, the writers associated with the Berlin School, whose contribution to the ensuing aesthetic of music should not be underestimated, considered themselves educators and critics more than philosophers. Based on Baumgarten's idea concerning the importance of resemblances in the poetic domain (see p. 174 above), Krause, for example, developed an argument, regarding the uniqueness of the aria, not very different from that of Smith: "No affect," writes Krause, "can be fully depicted in so few words as an aria contains; and the words are only there so that the composer has the opportunity to present the moving image on as many sides as necessary in order to arouse the whole affect, to which end the sentences and phrases of an aria must be well suited to repetition and fragmentation" (Hosler 1981: 107).

Krause aimed to create an analogy between the dynamic of a certain passion and the general dynamic of the da capo aria, implying thereby an imitative theory of sorts which he, however, does not explicate, let alone defend. The writers of the Berlin School, it seems, were still holding to "soft" versions of Pythagoreanism, modelled after Leibniz. Enlightened ideas of universal and natural language, of expressivity, of the uniqueness of each artistic domain, are fused, in their writings, with more archaic notions of world harmony. "Music is a universal language of nature which is only intelligible to harmonious souls. Its peculiar expressions, which it does not borrow from anywhere else, have a secret understanding with these souls," wrote Caspar Ruetz, a university-educated church musician from Lübeck.[50] Even Herder adopted a quasi-Pythagorean approach to music, though he could have easily accounted for music's affective powers by enlisting a theory of expression. Viewing the listening experience as an inward resonance to external sounds, Herder opted for typical German pantheistic mysticism, related to "cosmic music," while adopting an antiintellectual interpretation of Leibniz's "unconscious counting of the soul." His theory of cosmic resonance is related to both ideas, as well as to the upsurge of instrumental music, which Herder could not help but praise for its independence of external dictates.[51]

Judged from the British point of view, the attempts of the Berlin thinkers resemble those of their German compatriots in other aesthetic areas, for they, too, were more interested in norms than in the processes relating to the perception of their forms. They, too, were more concerned with beliefs and tastes than with faithful descriptions of the art, whether from a historical or aesthetic point of view. Their attempts, nonetheless, may be viewed as significant steps toward the Romantic ideal of "absolute music." Considering this development, a word should be said about Kant's treatment of music. Music, of course, could not achieve much encouragement from its categorization as "the art of beautiful sensations."[52] Emptied of conceptual content, music found itself naturally in the domain of "independent beauty." Kant wished to judge music in its "absolute" state, stripped of texts, dramas, and the like, in order to place music in his general classification of the fine arts. This also provided the ground for his view of music as representative of pure thought processes, a conception not unlike that of Smith. But whereas Smith viewed music's combinatory and varying procedures as "filling the whole capacity of the mind so as to leave no part of its attention vacant for thinking of anything else" (*AM*: 417-8), Kant viewed the play of thought excited by music as "the effect of more or less mechanical association," and on the whole, "more annoying to us than agreeable" (Kant 1952: 194-6). Indeed, once the mystical/mathematical element is removed from the perception of music, the absence of an alternative perceptual theory makes the philosophical status of music very low, indeed.

Kant, however, represents a clear exception in the history of music aesthetics in Germany, and his influence on later generations was, in this respect, limited. Instead, it was the Berlin writers who were decisive for eventual musical thought. Like Kant, Moritz, the teacher of Reichardt, Wackenroder, and Tieck, replaced "mimesis" with the aesthetics of "disinterestedness," employing it, however, as a step towards the Romantic conception of "organicism." His failure to entertain the possibility of measuring artistic perfection from inside art, however, left the field open to intuitive judgments and to the association of art with the ineffable—the unverbalizable (Wellek 1981: 209; Neubauer 1986: 180-1). Unlike the perceptualism of our British thinkers, this kind of approach to art drove Moritz's disciples to associate art with metaphysics. What is perhaps most revealing, as far as the disciples are concerned, is that they vested their ideas in literary discourse instead of philosophical ones, as Dahlhaus emphasized (Dahlhaus 1989, chapter 4), avoiding a debate about their ideas, to begin with. Through literature, they could affect a change in the system of categories of reception, while constituting a new cult—the cult of music. This new cult became associated with new cultural ideals manifested in the preference of loneliness over the communal, the fantastic over the realistic, the ineffable over the well defined. While the ineffability of music was already highly regarded by Luther, once it became a cultural ideal it turned the attempts to elucidate its meaning into a contradiction in terms. Consequently, a linguistic account of music could only rest on metaphors, analogies and examples. From Moritz to Schopenhauer the literature is packed with metaphors and analogies, as though causal relationships had never been "invented."[53]

When Nägeli tried to banish all traces of character from music, claiming that affections and scenes—the characteristic of the pictorial—should altogether be banned from the musical "play of form," and when Hoffman condemned all narrative programmatic and descriptive music as "unpoetic," it became abundantly clear that they had opted for an aesthetic molded after existent musical examples, replacing, in their own way, the former mimetic ideals.[54] Thus looked at from a broader perspective, the British and the German thinkers, despite their differences, seem to have related to the same *overall* development in the history of ideas, as Abrams argued long ago (Abrams 1953).

Of course, the *actual* connections between German and British writers had an immense influence on both. Still, the difference between them, as this book tries to show, is not a simple matter. Whereas the contribution of the German thinkers to major musical and cultural developments in the nineteenth century is well established, the historical importance of the British is less clear and less direct, and hence also far less known. The British, we claim, suggested nonetheless "the road that *was* finally taken," but only after many years of additional deliberation.[55]

Notes

1. These five names by no means constituted a school of thought. Padre Giovanni Battista Martini (1706-1784)—the illustrious "composer of composers" from Bo-

logna—wrote a treatise on counterpoint and his *Storia della musica*, which unfortunately did not go beyond Greek music, partook in the creation of musical historiography in the modern sense. Francesco Algarotti (1712-1764)—an international figure, a friend of Voltaire and Frederick the Great—is known for his "*Saggio sopra l'opera in musica*" (1755) which had a wide influence, particularly on the reform in opera (especially on Traetta and Gluck). His main idea—that in opera poetic ideas should govern the musical constituents (a continuation of Galilei's legacy)— was quoted by Webb in connection with opera. Benedetto Marcello (1686-1739)—composer, writer, and theorist (whose eight volumes of psalms were edited by Avison, among others) was acclaimed for his *Il teatro alla moda* (1720), a celebrated satire on contemporary opera. Esteban de Arteaga (1747-1799) was on close terms with Martini, and, like Algarotti and Marcello, his several extended treatises deal with the problem of the ideal union of poetry and music in opera. Unlike the latter, however, he was deeply engaged in aesthetic thought, of which he had considerable knowledge. Lodvico Antonio Muratori (1672-1750) is mentioned here because of his innovative suggestions regarding descriptive poetry, which should aspire to "a united form that is intellectually cogent," as Hagstrum put it (1958: 161), thereby conferring a new perspective on the *ut pictura poesis* discussions. Muratori was known in England and may have influenced some of the above mentioned writers.

2. For the centrality of Rameau's theoretical writings see Cynthia Verba's (1993) fine book on the subject. See also Shirlaw (1917/1969) and Neubauer (1986).
3. The Ptolemaic conception of the cosmos, as a group of concentric spheres arranged hierarchically in ascending perfection, had long given way to detailed knowledge of the actual movement of the heavenly bodies which "dislocated" Earth from the center. This, of course, had far-reaching implications for the idea of the centrality of man in the universe. Yet Earth, as one of the planets revolving around the sun in the solar system, was a purely mathematical problem, not only for Copernicus, but for Kepler as well. While the latter had discovered mathematical laws describing the movement of planets, Galileo discovered mathematical laws describing the movement of bodies on Earth. Newton brought Kepler and Galileo together, showing that Kepler's laws of planetary motion and Galileo's laws of terrestrial motion were two aspects of the same laws; all motion, in fact, could be described by the same mathematical formula. Accordingly, matter moves as if every particle attracted every other particle with a "force" proportional to the masses and inversely proportional to the square of the distance between them. This force was *universal*—universal gravitation.
4. Rousseau, as is well known, was a key figure in the "querrelle des Bouffons," which was part of the heated controversies concerning opera which took place throughout the eighteenth century (Lowinsky 1965). In France, these debates engrossed all kinds of circles—artistic, literary and philosophical. The debates centered, at first, on Lully, later on comic opera and finally around operatic reform. While the underlying issues of all of these debates was the standing of musical expression—whether autonomous or determined by the exigencies of the dramatic action—they tended to take the form of rivalry between French and Italian models and eventually between the French and the Italian approaches to music in general, not only to music drama.
5. With Aristotle, as we have seen, the notion of "imitation" has lost its Platonic meaning. Having denied the ontological status of numbers and the conception of the world as founded on a numerical structure, music could not even hold its special

place among the arts, that is, imitate the "harmony of the world." Nor could its beauty transcend that of the work of art. The "artistic," claimed Aristotle, invariably exists in the work of art itself, since it is the artist who shapes its form, having recognized the potentialities inherent in its materials. According to Aristotle, the "mimetic" powers, which partake in the cognitive process of analogy, are inherent to learning. Just as he has enlarged the realms of dramatic imitation to include the passions, Aristotle also introduced into music "harmonics" which copy emotional states. See Katz & Dahlhaus 1987 i: 229-31.

6. Dubos (1719/1978), vol i: 4-21. As chargé d'affaires to England in 1702, Abbé Dubos met with some of the members of the English sensationalist school, a fact which no doubt influenced his *Réflexions critiques sur la poésie et la peinture*, which appeared in 1719. For an interesting discussion on the subject see Verba (1993: 36-8). Later, Dubos himself influenced the English, especially after his treatise had been translated (in 1748, by Thomas Nugent).
7. Batteaux (1969: 368). Batteaux's *Lex Beaux arts réduits á un même principe*, was published in 1746 and was also known in English, especially through the anonymous "translation" bearing the title *The Polite Art,or a Dissertation on Poetry, Painting, Music, Architecture and Eloquence*, which was published two years later. The differences between the original and the plagiarized text are of interest and should be studied unto themselves.
8. Fontenelle's "*sonate, que me veux-tu?*," quoted by Rousseau in his *Dictionnaire* (article "Sonate"), became a slogan for the contenders in the eighteenth-century debate regarding the meaningfulness of instrumental music (see, for example, Twining's quotation, in Vol. II: 358).
9. Like the other *philosophes*, Rousseau was deeply influenced by Rameau's theory, though he distanced himself, from the start, from its "universal" assumptions. See Neubauer's discussion of the various stages in Rousseau's musical thought (1986: 89-102).
10. For an excellent and detailed discussion of Rameau's ideas on these subjects see Verba 1993, chap. 2. Also see Shirlaw (1969 chap. IX).
11. See our discussion of Condillac and relevant bibliography in chapter 4.
12. See Bernard 1980, Neubauer 1986, and Christensen 1985: 56; 1989.
13. See Fellows 1977: 93; for the influences on his musical aesthetics also see Gay 1969 ii; and Oliver 1947.
14. Boyé 1779; for an English translation of a selection from his treatise see Lippman 1986 i.
15. Chabanon 1785; for a discussion of his theories see Neubauer 1986: 170-1; and Maniates 1969.
16. For the connection between Mersenne and Rameau see Gruber 1970.
17. Rousseau's comparison appears in his *Essai sur l'origine des langues* (1753), in the chapters: "On melody" and "False analogy between colors and sounds." For an English translation see Rousseau 1966: 53-55; 61-64. François Jean de Chastellux's *Essai sur l'union de la poésie et de las musique*, was published in Paris in 1765.
18. There was hardly a period in operatic history, which was not engaged in warfare against the "nonrealistic" tendencies of the medium. It often took the form of rivalry between "realistic" vs. "sensual" principles. The former were invariably related to the exigencies of language and the latter to those of music. The debates over these subjects are contained in a vast body of literature to which librettists, composers, poets, philosophers, and many men of letters contributed. Whether related to the struggle for and against French and Italian opera, for and against opera seria, or

altogether, for and against opera as drama, attackers and defenders alike never lost sight of the fact that music was a more "idealistic" language than the rational discursive one. For a fine selection representing the entire range of the debates, see Weisstein (1969). For a discussion of representative views, see Katz (1986), chapter 2.

19. Unlike the religious wars of earlier times or ideological wars of later times, those of the eighteenth century centered less on principle issues. They centered primarily around the duel of Britain and France for colonies, trade and sea power, and on the struggle between Prussia and Austria for territory and military power in Central Europe. The French favored a divided Germany, that is, the separation of Austria and Prussia, for fear that a union between them would threaten the other European powers. The English were greatly impressed with the military strength of Prussia and tended to support Prussia in its fights against Austria. The French alliance with Austria was, however, never popular in France. The *philosophes*, for example, viewed the Austrians as a backward Catholic nation compared to the "free-thinking" Frederick of Prussia. These sentiments contributed eventually to the revolt against the Bourbons. Frederick, himself, was a man of letters and though he was contemptuous of the idea of the "divine rights of kings," he wished to safeguard the rights of the Brandenburgs; moreover, he appreciated the role played by the state. This was also the reason for his invasion of Silesia and its subsequent annexation to Prussia. The alliance between France and Austria divided the two camps into a northern progressive one against a southern conservative one, controlled by old-fashioned Catholic monarchies. In the final analysis, however, the wars of the so-called "Austrian Succession" (1740-8) and those of the "Seven Years" (1756-1763) centered on the control over an economic world which kept growing in power and importance. By the second half of the eighteenth century, the Prussian aggression of 1740 became not only legalized, but was viewed as a heroic act on the part of Frederick who successfully "defended" his state. From 1763 until his death in 1786, Frederick was indeed a man of peace, interested in culture and philosophy. Prussia, though, did not only survive in its enlarged dimensions, but began to glorify its army and consider its strength as a "mission." Not willing to trust anybody less capable than himself, Frederick died without having trained any successors. Twenty years later, Prussia was all but destroyed by Napoleon. It was eventually concluded that government by a "master-mind" working in lofty and isolated superiority was not a viable form of state under modern conditions. It is no coincidence, of course, that the majority of the debates, which tried to introduce novel ideas, took place in Prussia. Nonetheless, our short survey does not do justice to the complexity of the political and social forces that characterized the entire German scene. Nor does it explain why music flourished in particular in a cultural center like Vienna.

20. The difference between the above mentioned musical figures and the philosophical and the literary ones is apparent. Posterity proved that Bach remained no less central in the German pantheon than his followers, while Opitz and Wolff, for example much less so. Yet it was Mozart and Beethoven, who, in the first place, contributed to the very emergence of such a Pantheon, in which Bach subsequently found his honored place. As we shall later clarify, it was the *Frühromantiker*, more than their knowledgeable predecessors, who contributed to the building of such a pantheon.

21. By "code" we mean the sociological concept related to a set of "programmed instructions" directing social systems to respond to a recurrent set of problems in a particular way. Central to the concept is the adequacy of its description of the

processes of continuity and change (Katz 1985). Of course, our suggestion with regard to a prevailing "German code" is not solid enough to serve as a formalization of hidden directives of German history in the centuries under discussion, let alone to put forward its underlying structure and dynamics. Rather, our attempt is to be viewed as a preliminary explanation of the events that seem most pertinent to the handling of the problem under discussion.

22. Hermeneutics of the Holy Scriptures was indeed basic to the Lutheran tradition, which strove to reconcile the word of God with the newly established system of beliefs. Central to this tradition—basically established by Melanchton—was the emotional constituent, that is, the evocative reading of the text, which attempts to arouse in the reader the appropriate religious affect. This had an immediate implication for the developments under discussion.

23. This and the following quotations are from Luther's letter of 1530, to the composer Ludwig Senfl, asking for motets [F. A. Beck, *Dr. M. Luther's Gedanken über die Musik* (Berlin 1828); quoted in English in Fubini 1990: 143] Luther's view of the benevolence of music should be considered as a reform in itself. Even St. Augustine, among the church fathers, who surpassed all the other Roman Catholic authorities in his adoration of music, revealed reservations, precisely because music is able to divert the attention from the religious content to the sensuous pleasure. It is for such views, and more derogatory ones, that pure instrumental music hardly found its way into the Roman Catholic churches, as Fubini argues (ibid.).

24. Ibid. It was Melanchton's reform of the German educational system that put Luther's ideas, regarding musical education, into practice, already in the sixteenth century. Textbooks published in that century are exceptionally progressive, as Ian Fenlon maintains (1980). Among other things, the general musical education included singing for an hour a day. These tendencies were reinforced in the seventeenth century by people like Comenius, Praetorius, and Franke, in contradistinction to Locke in England, who declared that musical learning was a waste of time (see Ringer 1980). While the decreasing religious motivation seemed to endanger the high regard for musical education in Germany, rescue came from the special German readings of the French conception of individualism, which we discuss below. Thus Rousseau's conception of the role of musical education in the emotional development of the child was adopted by the Germans, especially by Bernhard Basedow, boosting the prevailing system while enlarging its scope.

25. Still, the great upsurge of German instrumental music associated with the church had to wait until the seventeenth century and kept growing with time.

26. Kircher's oeuvre is dominated by a vehement "quest of lost knowledge," as Godwin (1979) put it, mixing ancient and modern frames of thought. As far as music is concerned, it is revealed in activities as divergent as the invention of several acoustic devices and instruments, detailed descriptions of instruments long obsolete, notation of animal cries and of Jewish cantillation, and, above all, in his reintroduction of the Pythagorean beliefs. The latter stood in no contradiction to his developing of an elementary computer, serving as a "composing machine" (ibid.). His experiments in this regard should be considered as the beginning of *ars combinatoria* which engaged many a composer during the eighteenth century (Ratner 1970).

27. It should be emphasized, however, that Leibniz circumscribed his possible worlds by logic, including the law of sufficient reason. The variety of his worlds consisted accordingly of different states of affair, different "natural laws." This is exactly what

was so suggestive for aesthetic theory, which tried to apply Aristotle's probability to the fictive and figurative at large. Not only Baumgarten, but also Bodmer was attracted to this kind of thought (Wellek 1981: 147).

28. Each of the musical-rhetoricians had his model—a composer, whose work(s) stood as an ideal example for the "techné" which he elaborated upon. Burmeister's model was Lassus, Bernhard's was Schütz, and Mattheson's was Marcello. Canonization, in this regard, had a special meaning, since the music rhetoricians considered their models, however paradigmatic, as innovative.

29. In his essay "Norm and Form: The Stylistic Categories of Art History and their Origins in Renaissance Ideals" (Gombrich 1986: 81-98), Gombrich shows how historical categorization and awareness of stylistic differences have been intertwined with normative categories since Vitrivius and Vasari. Only toward our own century, argues Gombrich, did they lose their derogatory connotations. Gombrich further argues that the two major stylistic categories, the classical and the nonclassical, underlie all the others (Romanesque, Gothic, Renaissance, Mannerist, Baroque, and Romantic) and as such were basic even to Wölfflin and Curt Sachs. Thinkers in the eighteenth century relied, in this respect, on Renaissance typologies, pulling them in the direction of the historical-descriptive (like the English) or reemploying them for cultural-normative needs (like the Germans). Winckelman's adherence to the classical ideals may be thus interpreted as an attempt to purge German art from its "gothic" orientation through the Italian norms which derided it, giving art the necessary classical twist it needed.

30. Mendelssohn's ideas in this regard appear in his *Über die Empfindungen* and his *Betrachtungen über die Quellen und die Verbindungen der schönen Wissenschaften.* (Mendelssohn 1971 i: 43-119; 165-190.) For a discussion of Mendelssohn's position in the development of German aesthetics see Kristeller 1965: 217-220. The following remarks are partially based on Kristeller's discussion.

31. In a rather Platonic way, Mendelssohn claimed, that the "göttliche Vollkommenheit" as conceived by reason, is always above sensual beauty, which is inferior because of the ease by which it suggests its perfection to the perceiving mind (Mendelssohn 1971 i: 59-60).

32. Mendelssohn collated many more authors, among whom were Dubos (in connection with the distinction between arbitrary and natural signs; Wellek 1981: 148), Lord Kames, Batteux and others; as such he may be considered a man who had an overview of the subject.

33. This famous expression is taken from his passage on the Laocoon, located in his *Gedanken über die Nachahmung der griechischen Werke* (1755). For a partial translation see Nisbet 1985: 29-54.

34. Nisbet argues that Lessing's position resulted in normative views even stricter than those of Winckelmann, especially with regard to the visual arts. Here he relies on Gombrich who stated that the *Laocoon* "is not so much a book about as against the visual arts." See Nisbet 1985: 10 and Gombrich 1957.

35. Lessing's ideas on this subject appear in the *Hamburgische Dramaturgie* (Lessing 1958: 103-11), a collection of his critical essays. The occasion for expressing these ideas on music was Agricola's *Semiramide*, the instrumental parts of which he did praise. Basically relying on his colleague—the music critic Scheibe—Lessing remained faithful to his conception of the limits of artistic media, allowing music to speak its own language only insofar as it does not encroach into foreign domains. The juxtaposing, to say nothing of mixing, of contrasting affects within one and the same movement, is aesthetically unjustifiable, maintained Lessing, since music, by

definition, lacks the power to reasonably account for it. For a fine discussion of Lessing's extracts on the subject see Hosler 1981: 137-141.

36. This is strange indeed, if one takes into account the fact that already in 1741 Schlegel (Johann Elias) argued, that not the resemblance to reality is the core of artistic imitation but the effect, the emotional impact (Wellek 1981: 146). He, in turn, only emphasized what he might have found in Baumgarten's writings.
37. Webb, for example, could afford to criticize classical prosody, opting for artistic norms more akin to the nature of the English language and to an alternative artistic desideratum.
38. Nisbet 1985: 6; Wellek argues that in the case of Bodmer, who together with Breitinger are responsible for the importation of some of the English thought into Germany, this "middle-classishness" warped his imaginative insight (Wellek 1955: 148).
39. Moses Mendelssohn, for example, writes: "Der äußere Mensch ist versorgt. Wir können uns der Mittel bedienen, die in allen Jahrhunderten sind erfunden worden. Die Natur bleibt immer eben dieselbe. Allein der innere Mensch ist unbebaut. In jedem Jahrhunderte betreten andre Menschen die Scene des Lebens. Sie müssen alle sich mit würdigen Gedanken beschäftigen, und die marternden Zweifel aus ihrer Brust verbannen." *Über die Empfindungen*, Mendelssohn 1971 i: 63)
40. For a discussion of Hamman's main ideas see Wellek 1981: 178-181; Nisbet 1985: 11-15.
41. Herder was by no means the first to pose competing, expressive ideals against the dominating classical ones. Bodmer is important in this regard, for he was the discoverer of German medieval literature (*Parzival, Nibelungenlied,* the great *Minnesänger* manuscript), Dante, and others. Heinrich Wilhelm Gerstenberg (1737-1823) was also very influential in this respect (Wellek 1981: 147; 176-8). Expressivism, as a style, was paradoxically introduced by Winckelmann himself through his sensuous, evocative literary style.
42. The German conception of the beautiful and the sublime was basically imported—by Mendelssohn, Lessing and Kant—from Burke's seminal inquiry on the subject. As we have pointed out, Burke bestowed upon the concepts an empiricist analysis, in the spirit of Hume's theory of mind (See chap. 4 above). According to Boulton, Lessing found in the *Enquiry* confirmation of ideas he had developed independently; he made no attempt to analyze its philosophical tenets, of which he basically disapproved (Burke 1958: cxxiii). Still adhering to a mimetic theory of poetry, Lessing was in fact unwilling to cope with the psychological observations of the English writer. Kant was different; he acknowledged Burke's central importance to the empirical study of aesthetics, which he considered basic to all philosophical inquiries on aesthetic judgment. Unlike Burke, Kant, as noted above, went beyond the principle of self-preservation, subjugating it to higher intellectual and moral principles (Burke 1958: cxxvi).
43. Schiller's treatise *Über naive und sentimentalische Dichtung* was first published in 1795-6 and is considered the most important theory of poetics to appear in Germany after the *Laocoon.*
44. See Wellek's excellent discussion of Schiller's theory (1981: 227-55), which he juxtaposed to that of Kant. See also Nisbet 1985: 20-24. Chytry's instructive discussion of Schiller's ideas furthers the analysis of Schiller's debt to Kant and his distancing from him (Chytry 1989: 70-105). The more prominent context in his study, however, is the German concept of the "Aesthetic State," which owes much

to Shaftesbury, particularly in the case of Schiller himself, as Chytry convincingly demonstrates.

45. Scheibe's attack on Bach's turgid (*schwülstig*) and confused style is well known, and so is Birnbaum's rebuttal (David & Mendel 1966: 237-252); it was indeed the "singing" quality of music which both Mattheson and Scheibe looked for, each in his own way. For a discussion of Mattheson's special stance in this regard see Katz & Dahlhaus (1990 iii: 63-66).
46. Mattheson 1981: 482-4; Willheim 1963: 162. In chapter 3 we dealt with the process of the "rhetorization of music" in the high baroque style. Yet rhetorizing music, we believe, did not stop with that style; the classical style reveals additional rhetorical aspiration both with regard to the syntactical as well as the structural elements of music.
47. The phrase relates to Constanze Mozart's utterance—about the fugue being the most "artistic and beautiful of all musical forms"—which she expressed at the time when the craze for the fugue in Vienna, encouraged by Von Swieten, was at its height (King 1955: 170).
48. The enlightened school in Berlin was, of course, a direct outcome of the enlightened climate that dominated Prussia at the time. See fn. 19 above. Mendelssohn and Lessing were among its illustrious members as well as the "founders" of the Lied as an important musical genre.
49. This quotation, from Christian Gottfried Krause's *Von der musikalischen Poesie* (1753) is taken from Hosler's faithful account of Baumgarten's music-theorist disciple (1981: 96). Our following discussion is based on her account.
50. See Hosler 1981: 126. Ruetz (1708-1755) expressed his ideas on music in his "Sendschreiben eines Freundes an den andern über einige Ausdrücke des Herrn Batteux von der Musik" which were published in Marpurg's journal (*Historisch-kritische Beyträge zur Aufnahme der Musik*; 1754-60) in 1755. If Ruetz articulated the reaction of most German church musicians to the French neoclassic doctrines, as Hosler argues, our basic claim finds an additional important hinge.
51. Herder's ideas on the subject, discussed by Neubauer (1986: 159-163), were first developed in his *Critisches Wäldchen*, of 1769 and later in the *Kalligone* (1800) in which he formulated the above view. Herder's attempt, in the earlier treatise, to discuss the phenomenon of sound from a sensual point of view is of interest in comparison with the perceptualism of the British group. Unable to bridge relations and proportions of sound, on the one hand, and its psychological impact on the other, Herder, according to Neubauer, opted for an anti-Rameauian position holding that it is simple sound, devoid of overtones, which constitutes the aesthetic value of music. Clearly influenced by Rousseau, Herder's "*monadology*" of sound (as he coined it), misses, of course, Webb's basic observation that only *succession* of impressions (irrespective of their harmonic load) can account for the aesthetics of music.
52. For Kant's text and a short analysis see Katz & Dahlhaus (1987 i: 293-314). See also Neubauer (1986: 182-192).
53. Of course, people like Kirnberger and Koch did engage in a discussion of "causal" relationships, yet only as far as compositional "preference rules" are concerned. Koch's ideas about the musical potential of a period, for example, were only indirectly related to epistemological and perceptual queries; as such, they hardly entered into the aesthetic discourse. Koch and Kirnberger's attempts, considered by Bent (1980) as pioneering steps in musical analysis, were, in fact, a continuation of the musico-rhetorical tradition, while believing music to emanate its own terms and

criteria. Moreover, though addressing new ideals—primarily the ideal of organicity (Broyles 1980), they took more from music aesthetics than they contributed to it.

54. Nägeli (1828) took issue in this regard with Körner who developed the idea of the centrality of "moral character" to musical expression (1795). See Dahlhaus 1989 on the subject (42-45; 65-68).

55. Still, there are good reasons for the rereading of the Germans with the British treatises in mind (especially those which were translated into German like Webb's, Beattie's, Burke's and others'). Careful examination may perhaps reveal hidden paths through which major seminal ideas were transmitted, leading to our own century.

7

Postscript: Sense and Meaning Interlocked

The new trends in seventeenth- and eighteenth-century art and theory reveal a fundamental change in the conception of art and art making. As we have seen, the general theory of mimesis was carefully scrutinized and examined in its detailed applications. The process engendered theoretical crystallization through "corrections" and refinements of old conceptions, giving rise at the same time to new queries, which in their turn, yielded new explications. Moreover, the relationships among the arts underwent changes in the course of this process, revealing new aspirations with regard to each of them.

The picture that emerged from all these activities vividly portrayed the interdependence and intimacy which exist among art theory, art making, and the aspirations of those who take an active part in the fostering of both art and its understanding. It must have become more and more clear that art embodies the very conceptions of its modes of representation. Moreover, by objectifying the spiritual in sensible matter, art seemed to create an enduring unity of a self-contained form.

The notion that art embodies understanding and that particular understandings are revealed in works of art is propitiously upheld nowadays in many circles—literary, philosophical, and others—despite the communicative difficulties that were raised by modern art. Such a proclamation, however, is not as simple as it may sound, for it actually involved and rested upon the accumulated understanding with regard to the working of symbols and the function of symbolic forms not only in the arts but as part of a total theory of knowledge.

It is interesting, though not accidental, that the understanding regarding the centrality of symbolic forms in man's efforts to conceptualize himself, his actions, and the world he lives in, should have found a natural habitat for the threshing out of the problems and issues it involves within the precincts of art. As a testing ground embodying intricate conceptual ideas in perceptible images, the sensible world of the arts gave the interaction between the two a saliency wanting in other domains. Of course, in all of its transformations, art never claimed to explain Nature—to unveil the substance of its being—but rather to mirror, express, or highlight it in one form or another. That the "one form or another" reflects man's conception of the world and not the real world, gradually dawned upon those who combined the two—conception and world—in their artistic creations. Indeed it is this awareness—i.e., that the world of art does not reproduce the real world but creates its own universe, a self-revelation which it communicates through symbolic forms—that gained momentum in the eighteenth century. Although most salient in the arts, it found expression elsewhere as well. This is particularly true of the notion that symbolic forms contain within themselves the very conceptions which give rise to them and which they serve to represent, generating unexpected interactions they stimulate between mind and matter. "Conceptions" thus conceived, clearly function not only as means of representing the human spirit, but as substantial elements of the spirit itself.

This implied an entire revolution in the theory of knowledge, effecting shifts in epistemology as well as in subject matter. In fact with the realization that it is impossible to sever the human spirit from its products, including man's conception of the world, man-made culture replaced God's creation, and the "how" of man's creation replaced the blueprint of God's vision.

Though the eighteenth century left many questions concerning culture unanswered, it was the century which presented the phenomenon of culture as a key element in its vigorous scientific agenda. The initial stages of this scientific activity entailed the reexamination of the role and function of signs and symbols. Within the latter re-production and production fused into one, replacing temporal powers of referentiality by a continuous generative force creating and changing a humanly conceived world.

We had occasion to point out that the rise of modern art, especially those trends that cast aside all figurative representation, is inconceivable without the deliberations of the kind discussed in this volume. Modern aesthetics is likewise indebted to these deliberations whether it focuses on form, intentions, or differences among symbolic systems. And last but not least, by considering the working of the mind as it processes information, these deliberations contributed as well to the present ongoing concern with cognition, not only in art and art theories. Indeed, it is fair to claim that the way we think about the arts can teach us something about the way we think generally. Specifically, the changed conceptualization of the arts—the move from mimesis to world making—has obviously affected cognitive theory, though it has, on the whole, been understated.

Cognitive studies are concerned, generally speaking, with the manner in which the mind processes information and structures coherence. To move from process to coherence, we are told, requires some form of construction or representation, based on relationships among the particular particles which are processed. In other words, process is influenced by that which is being processed. Yet coherence, structured as it may be, does not in itself carry meaning; in other words, coherence makes sense but has no meaning. To turn such meaningful cohesions into cohesive meanings requires symbolization.

In recounting the contribution of the various sciences to "the mind's new science," Howard Gardner (1985) enlisted the entire gamut from philosophy to neurophysiology, via anthropology, linguistics, psychology, and artificial intelligence. It is to Gardner's credit that he viewed philosophy as central. It is, after all, the struggles between the rationalists and the empiricists, the debates concerning the innate vs. the learned, and the relations between the "given world" and the "possible worlds" that emphasized the distinction between the inner activities of the mind and its interactions with that which is external to it. The very debates that yielded such distinctions also pointed to the difficulties in sustaining them theoretically, not to mention separating them empirically. Gardner (though he is not the only one) neglected, in this context, the lessons that could be learned from the ways the arts went about creating and developing coherent worlds and the ways they were conceptualized as doing so.[1] As we have seen, this is an extended narrative in which music played a central role. In what follows, some of the *key* factors which provided crucial links in the scaffolding of this extended narrative will be weaved together in order to present a single graspable picture of our major claim.

We began our narrative of the move from the conception of art as *techné* to a conception of art that emphasized creativity as a common denominator in the different arts in the late Renaissance. Painting was central to this development. In the course of the Renaissance the artist was more and more expected to render reality *as if* it were taking place before his eyes, eternalizing, as it were, the fleeting moment. From this point on, the role of true art was no longer solely to imitate nature as such, that is, to produce appearances in the Platonic sense, but to present nature as *perceived* by the artist. The epistemological implications of this shift were, however, fully grasped only in the eighteenth century. By then, Baumgarten, in his investigations into the aesthetic, could maintain, categorically and unequivocally, that nature neither explains nor dictates our understanding and construction of art. He did not mean to imply that creating art was an activity void of constraints, but rather that the constraints are those of human perception striving for coherence. Although Baumgarten relegated the arts to an "unlogicized lower cognitive faculty" characterized by "fusion" and "continuity," it nevertheless aspired, he claimed, to the cognitive notions of order and unity.

If Baumgarten emphasized the centrality of inner consistencies achieved through associations, resemblances, internal references and the like, he did so because he believed that these were necessary to make a work of art coherent, providing a parallel to the upper faculty's principle of "sufficient reason." According to Baumgarten, mental processes are essential ingredients of perception *itself*, and the coherence of a work of art reflects the working of the mind. One might say that Baumgarten was more concerned with "form being formed," to use a favorite phrase of aestheticians, than with the subject or subjects communicated by the work of art.

In Baumgarten's times—for he was not the only one, as this book has shown—it became clear that art resides in its making—in the ways in which it is constructed. (For all its novelty at the time, let us recall that Aristotle identified the arts with the capacity "to make," invoking "a true course of reasoning, the origin of which is in the maker and not in the thing made.") Of course artists were busy all along with the construction of works of art; moreover, some even wrote treatises on "how to go about it." But these were mainly related to the knowledge of the métier and not to queries concerning the constraints of mental faculties. It is the growing awareness of the latter that we wished to stress.

The concept of perception in relation to the arts, even when it was still barely developed theoretically—was linked to another cognitive concept, the imagination. The ability to perceive a "solution"—a sense of some whole—was deemed necessary in order for perception to complete its own organizational task. Interestingly, the more this ability was assigned a part in perception not only with regard to the arts, the more it caused the "real world" to recede from human grasp. From Leibniz, Shaftesbury and onwards, attempts were made to fathom the power of imagination as a constructive faculty creating its own referential systems. The world was not ignored thereby, it only seemed more out of reach as an entity independent of its human formulations. In due course, the world as perceived came to be viewed as a kind of "representation," circumscribed by the boundaries of human cognition, as Kant would soon maintain. Paradoxically, the tangible world received a fictive status, and this fictive status acquired an ontological standing.

Relegating the arts to the realm of perception led to significant inquiries into the correspondences and differences among them. Even if Kristeller tells us that the "fine arts" were thought to comprise a distinct group of activities only in the eighteenth century, this is better seen, as he himself recognized, as the culmination of seventeenth century attempts to apply a scientific method to all areas of human knowledge and activity. Ironically, perhaps, art failed to meet the tests of cumulative progress and lawfulness. Just as the failure to deal with the lawfulness of human perception and irrationality has led to the separation of art from science, so the pursuit of perception has led to the separation of the arts from each other. Gradually, the idea that the several arts were different treatments of essentially the same subjects yielded to the idea that different artistic media dictate the choice of compatible subject matter. The eighteenth century is replete with such illuminating treatises of which the well-known *Laocoon* is but one.

As long as attention centered on subject matter—on that which was being imitated—the arts felt free to borrow from each other their modes of symbolization. This was why such importance was attached to the *ut pictura poesis,* which rested on the premise that both poetry and painting—and the other arts by extension—were similarly related to the world. As Hagstrum has shown, poetry at times became iconic, even tableau oriented, while painting became historical, dramatic, and allegorical. For a long time, it was "mimesis" through which the problem of reference was treated and the notion of symbols was handled. These were central for explaining the communicative aspects of the arts. By the eighteenth century, when the reality of the imitated world became suspect, it was agreed that the modes of symbolization themselves dictate what is being symbolized, thus reiterating, from a different angle, the statement that art resides in its making. It enabled a Twining, for example, toward the end of the century, to claim that modes of symbolization must be refashioned if one art is to function in ways indigenous to another.

To repeat, the theory of art freed itself only gradually from the constraints of the concept of imitation. Before the arts could be defined as symbolic languages through their own organization and symbolization, namely through their "sensuous, form-giving character," to quote Cassirer, one had first to realize that "images" of whatever kind were insufficient for the understanding. "Images" have to be turned into symbols by the very intellect they address, using the content of perception and experience. Thus, symbols are not only made, but disclose themselves through their own making; they are in fact the cultural forms of human activities. Looked at in this way, one might say that awareness of the "interpretative capacity" of the arts preceded other cultural forms, since the sensible worlds of the arts were first recognized as man-made. By rendering perceptible images into intricate conceptual ideas, art lent the interaction between the two a kind of saliency wanting in other domains.

By the time Kant added aesthetics (as a special branch of critical philosophy) to the Greek division of philosophy into physics, ethics, and logic, a philosophy of culture, along the lines mentioned, was already well on its way. Without much variation, it was applied to other forms of thinking—judging, knowing, understanding, and even feeling, by which the human mind attempts to conceive the universe as a whole. Moreover, it had become increasingly clear that the unities which were thus created in the various cultural domains could not be conceived in the manner of systematic metaphysics as indivisible entities, but that they had to be understood and defined in functional terms, such as relations, operations, and actions. Symbolic representations, in other words, cannot be abstracted from their various manifestations as self-contained, isolated existences.

Indeed, the Cartesian insistence on the deductive method as a sole path to knowledge had by then already been challenged in many ways for its suppression of the imagination. Vico's *verum* had altogether turned Descartes's mathematical reasoning into an "invention of a symbolic system which men can logically guarantee only because they themselves have made it." The realization that knowledge is neither Platonic nor Cartesian, but a social process traceable through the evolution of symbols, became amply clear.

Let us delineate the picture which emerges from what was said so far. As we have tried to show, focusing on creativity highlighted the part of the artist and of the beholder in bringing about the work of art. Implicating the artist and the beholder highlighted process—process in the cognitive sense—which is intimately related to the artist's perception rather than to procedures of craftsmanship, though the two do not necessarily exclude each other. Calling attention to perception, however, did not suffice to account for aesthetic qualities. Only when the particles are organically interrelated does perception acquire its proper standing in the arts. Coherence was thus introduced into the discourse, and related to process. With the concept of coherence, the constructive elements of art gained new saliency, and with this new understanding of constructive procedures, it has also become clear that each art generates its own mode of symbolization.

A separation was then set between sense making and meaning. While sense making highlighted process and coherence, meaning was grafted onto the constructive elements that create coherence, which are necessary, of course, for its mode of symbolization. Sense making was increasingly related to the innate operations of the mind, while meaning became more and more identified with aspects of culture, conventions and their likes. The separation between sense making and meaning notwithstanding, they remained intimately connected through the constructive elements.

That all of this should call special attention to the art without semantics should not come as a surprise. The constructive always loomed large in music since no reality could easily and unequivocally be related to it; in short, it has no ostensible referent outside itself, and yet it allows for great specificity. In the universe of sound, the sense data permit a very high degree of articulation, that is, the assignment of tones to specific places and functions along different dimensions within the totality of a musical system. No wonder music was very early recognized as one of the most potent foci of human intelligence which was, also, coupled with abstract structures; hence its continuous relatedness to various theories of numbers. Plato, who was ready to banish the arts altogether from his ideal State because of their deceptive nature, held music in special regard for, he claimed, it embodied the thought of the Demiourgos who saw fit to structure the universe on mathematical relationships. Music was *likewise* constructed, as we have learned from Pythagoras, and its sounds made the order of the world audible. Strange as it may seem, it is worth noting that these very premises enabled a scientist like Kepler to present his three planetary laws, using harmonic relations as his point of departure. It should be noted that it was music which could be of aid in the unveiling of The Thought of all thoughts.

Without further elaborating the far-reaching ramifications of the so-called "harmony of the spheres," it is also worth noting that Platonism and Neo-Platonism have always seen man in the likeness of God, granting his intellect a godlike quality. If the sensible world was the very manifestation of a perfect prime order, man-made music must potentially reveal that very same perfection. But whereas the understanding of order in terms of coherence had to wait for explications, the harmony created through particles tuned to each other, was immediately tangible. This is what prompted a Leibniz, for example, to state that listening to music is "the secret calculation of the soul unconscious of its calculations." With a statement like this, Leibniz revealed his awareness of a process stimulated by music; unfolding in mental time, connecting process to structure, to coherence.

Despite such unique characteristics, or perhaps because of them, the status of music as an art remained long unclear. Although process, structure and coherence surfaced early, it lacked a persuasive treatment of meaning. As long as "mimesis" reigned supreme, music was in fact considered lacking in cognitive standing, which goes a long way toward explaining the reluctance to sever music from language even in the eighteenth century. How is it then that not only did music gain in status toward the end of the century, but managed to arouse the envy of all of the arts in the nineteenth century?

On the basis of our discussions of the transformations that have taken place in theories of art, related as they were to the growing realization that man himself creates the worlds that are within his grasp, it is not difficult to guess some of the answers. Music contributed to, no less than it benefitted from, the growing awareness of a man-made world. Interestingly, while attempts to apply scientific method to all areas of knowledge gave music an advantage due to its seeming mathematical base, it was within the precincts of music that this base, as an explanation of the art, was challenged. Descartes had to admit that musical norms have a standing of their own and do not readily submit to calculated regularity. If Rameau succeeded for a while in persuading his readers that music functions in line with some kind of regularity, it was soon challenged by people like D'Alembert, Rousseau, and others. The latter were aware, probably more than Descartes, that music integrates cultural choices within its constructive elements. Indeed, while the propensities of the mind circumscribe the possible choices, choices may nevertheless vary. Such choices, however, become an integral part of the sense formation of the various musical systems, providing the grammatical base for musical relations, which circumscribe, in turn, the *latitude* of stylistic changes—not the changes themselves—retaining thereby their co-generic features.

Nonetheless, Rameau did reveal that music, at least the one he accounted for, is governed by a cohesive harmonic system. Even if his attempt to relate it to the Platonic theory of forms has failed, he has nevertheless shown that music had its own referential system (albeit a probabilistic one), that is, tonal grammar. Relations, hierarchy, and functions characterize this system, guiding the unfolding of at least that music with which he was concerned in significant ways. While the unfolding of music is never free and invariably calls for some kind of desired or assumed order, the development of Western music, in particular, was accompanied by a striving toward more cohesive and more encompassing musical frames of reference. Max Weber was right in claiming that the rational base he attributed to Western culture was best exemplified through its music.

Weber used the problem of musical intonation to make his point, but the point may be made just as well by reference to other musical frames. The history of Western music leads from the superimposition of external forms on musical materials, to musical forms emerging from within; from "melody types" through the modal system, scales and the unifying thorough bass, to tonal harmony; from a regular indivisible beat through simple mensuration to complicated relations between meter and rhythm; from the unfolding of single lines to complex relations among them. All these and more, found expression in Western musical theory. It should, however, be noted that we haven't yet mentioned even the most basic metaphors that accompanied all these developments from the start, beginning with simple binaries like high and low, through more complicated ones under the general heading of the so called, *via natura* vs. *via artis* to music's basic semblance—that of movement. Those were necessary for the establishment of "identities" in music.

The moment in which Rameau entered the discussion was indeed a moment of crystallization for these musical frames. Although he concentrated on tonal harmony, it should not be overlooked that he called attention to one of the most crucial aspects that govern the listening to music—i.e. expectation. The role of memory in the perception of music came to light already in the fourth century B.C., in Aristoxenus' pronouncement that music consists of "successive productions." "We must perceive the sound that is present and remember that which is past," says Aristoxenus, describing what he calls "the intellectual process of the apprehension of music" (Strunk 1981 vol i: 30). But the role of prediction became manifestly clear only by the time of Rameau.

However, awareness of expectation in music was advanced above all through the longstanding relationship between music and rhetoric. While many associated musical rhetoric with musical structure in a manner akin to language—phrases, sentences, figures, and larger units—it is more appropriate to define its task in terms of the logic of musical unfolding. In this sense, rhetorical figures were enlisted to highlight combinatorial rules and affective musical procedures. From Burmeister via Bernhard to Mattheson, both expectation and memory were accounted for through the employment of rhetoric, sometimes in convoluted ways. It is only toward the middle of the eighteenth century that dissatisfaction with rhetoric emerged.

It was Adam Smith, the man who put trust in an invisible hand which runs the economy, who firmly linked both musical memory and musical expectations to *learned* referential systems. He understood that only with the aid of these do processes turn into coherent structures, into forms, or "products." By that time, the idea had already dawned that a distinction must be made between innate propensities, or what Chomsky calls "competence," and the systems and worlds to which they give rise and circumscribe. This understanding guided Sulzer, for example, when he called for the separation of "universal rules of regularity" from the musical culture and styles in which they are manifested.

Attention to the cognitive building blocks that construct musical coherence finally brought new insights to the problem of meaning. The very process that led to the crystallization of referential systems for the constructive aspect of the art, also gave rise to referential systems for its representations. From the start, metaphorical uses accompanied the construction of music. When the floodgates opened, there followed a cascade of old and new signs, symbols, metaphors, and the like, not to mention the allusion and illusion enabled thereby. We wish to remind the reader of that impressive catalog of musical coinage that emerged in the course of the seventeenth and eighteenth centuries. In opera, this coinage functioned within a fictive world, which granted passions not only a delineated musical physiognomy but a psychological presence as well. These passions, of course, were superimposed, as it were, on the logic of musical structure. In retrospect, it seems that a stroke of genius befell Vincenzo Galilei and his friends when they spoke of "moving the passions *of the mind*" (our emphasis). Yet it remained for Descartes and his followers in musical thought to delve into the actual workings of the mind in this connection. They revealed the centrality of habitual associations culminating in what Nelson Goodman calls modes of "metaphorical possessions." Of course, membership in a culture is a prerequisite of the coding and the decoding of such meanings.

While musical meaning was (and still is) considered by many a mere appendage and a mostly irrelevant intellectual exercise, we have seen that toward the end of the eighteenth century, Daniel Webb succeeded in linking musical meaning itself to that which was long considered unique to music, that is, the semblance of movement. It is the latter, we learned from Webb, which makes possible the *externalization* of the cognitive process which music involves; it enables the pursuing of a pattern which the mind can retrace. Yet the traceable pattern is already vested with meaning so that the experiential is intimately wedded to what the mind recognizes.

Towards the end of the eighteenth century, in the discourse of some philosophically inclined insightful men of letters, an understanding was effected concerning the match between meaningful cohesion and cohesive meaning. The linkage between musical referential systems that guided musical construction and referential systems that guided music's representational signs acquired one of its most brilliant expressions in the high Classical style. Thus, Walter Pater in the nineteenth century stated that in music "the end is not distinct from the means, the form from the matter, the subject from the expression," that "they inhere in and completely saturate each other," and claimed that all arts aspired to that condition—"the condition of music."

Our main claim in this study is that the historical perspective concerning the ways the arts went about creating and developing coherent worlds is not only relevant, but may significantly contribute to contemporary discussions of the workings of the mind. As we tried to show, music played a unique role in the development that ushered in "the cognitive turn." Not unlike Kepler, *mutatis mutandis*, we too believe that as a "well-made world," music may provide a suitable point of departure for the examination of process and coherence in other worlds. It may have been music which provided the most profound theoretical insights concerning the relationship between the innate and the learned. Through music, the relationship between the permanent propensities of the mind and its permeable cultural representations became defined in ways more transparent and illuminating than they did in worlds less well made. Moreover, the intervening stages which bridge propensities and representations, calling attention to the *non*-direct relationship between them, also surfaced early in music. Music, above all, succeeded in proving that sense formation without predication is possible—thereby giving epistemology an early push toward cognition.

Notes

1. Gardner, however, is the last to be accused of not paying attention in general to the arts, as many of his other publications and activities attest.

Bibliography

Abbate, Carolyn. (1991). *Unsung Voices: Opera and Musical Narrative in the Nineteenth Century*, Princeton: Princeton University Press.

Abrams, Meyer H. (1953). *The Mirror and the Lamp: Romantic Theory and the Critical Tradition*, New York: Oxford University Press.

Ackerman, James S. (1961). "Science and Visual Arts," *Seventeenth-Century Science and the Arts*, edited by H.H. Rhys, Princeton: Princeton University Press: 63-90.

Adams, Robert M. (1971). "Where Do Our Ideas Come From? — Descartes vs. Locke," *Innate Ideas*, edited by S. P. Stich, Berkeley: University of California Press: 71-87.

Addison, Joseph. (1914). *Miscellaneous Works,* edited by A.C. Guthkelch. London: Bell.

Alanbrook, Wye J. (1983). *Rhythmic Gesture in Mozart: Le Nozze di Figaro & Don Giovanni*, Chicago: University of Chicago Press.

Alpers, Svetlana. (1961). "*Exphrasis* and Aesthetic Attitudes in Vasri's Lives," *Journal of the Warburg and Courtold Institutes* 24: 190-215.

________. (1983). *The Art of Describing: Dutch Art in the Seventeenth Century*, Chicago: The University of Chicago Press.

Alter, Robert (1967). "Fielding and the Uses of Styles," in *Twentieth-Century Interpretation of Tom Jones,* edited by M.C. Ballesin, New York: Prentice Hall.

Ambros, W. A. (1891). *Geschichte der Musik* vol. iii, Leipzig: Reprografischer Nachdruck.

Amman, Peter J. (1967). "The Musical Theory and Philosophy of Robert Fludd," *Journal of the Warburg and Courtold Institutes* 30: 198-227.

Anthony, James R.. [1978] (1997). *French Baroque Music from Beujoyeaulx to Rameau,* revised and explanded edition, Portland: Amadeus Press.

Apel, Willy. (1958). *Gregorian Chant*, London: Burns & Oates.

Arnheim, Rudolf. [1957] (1969). *Visual Thinking*, Berkeley: University of California Press.

________. (1974). *Art and Visual Perception*, Berkely: University of California Press.

________. (1984). "Perceptual Dynamics in Musical Expression," *Musical Quarterly* 70: 295-309.

Arnold, F. T. [1931] (1965). *The Art of Accompaniment from a Thorough Bass* in 2 vols. London, New York: Dover Publications.

Auerbach, Erich. [1946] (1953). *Mimesis, The Representation of Reality in Western Literature*, translated W. R. Trask, Princeton: Princeton University Press.

Avison, Charles. [1753] (1967). *An Essay on Musical Expression*, New York: Broude Brothers.

Bacon, Francis. [1627, 1859] (1963). "Sylva Sylvarum or A Natural History," *The Work of Francis Bacon* vol. ii, Stuttgart: Friedrich Fromman Verlag: 339-672.

________. [1605, 1627] (1960). *The Advancement of Learning and New Atlantis*, London: Oxford University Press.

Bakhtin, Mikhail. (1983). *The Dialectical Imagination: Four Essays*, edited by A Shukman, Austin: Texas University Press.

Balzano, G. J. (1982). "The Pitch Set as a Level of Description for Studying Musical Pitch Perception," in *Music, Mind and Brain*, edited by M. Clynes, New York: Plenum.

Barasch, Moshe. (1985). *Theories of Art from Plato to Winckelmann*, New York: New York University Press.

Barford, Philip. (1965). *The Keyboard Music of C.P.E. Bach*, London: Barrie and Rockliff.

Barry, Kevin. (1987). *Language, Music and the Sign: A Study in Aesthetics, Poetics and Poetic Practice from Collins to Colleridge*, Cambridge: Cambridge University Press.

Barthes, Roland. (1974). *S/Z*, translated by R. Howard, New York: Hill and Wang.

________. (1977). "Introduction to the Structural Analysis of Narratives," *Image-Music-Text*, London: Fontana.

Basilius, H.A. (1981). "Thomas Mann's Use of Musical Structure and Techniques in Tonio Kröger," *Literature and Music*, edited by N. A. Cluck, Provo: Brigham Young University Press: 153-174.

Batteaux, Charles. [1746] (1969). *Les Beaux arts réduits à un même principe*, Geneva: Slatkine Reprints.

Baumgarten, Alexander G. [1735] (1954). *Reflections on Poetry*, (with the original Latin text) translated and edited by K. Aschenbrenner and W.B. Holther, Berkeley: University of California Press.

Bayly, Anslem. (1778). *The Alliance of Music, Poetry and Oratory (Under the Head of Poetry)*, London.

Beattie, James. [1760](1778). *Essays on Poetry and Music as the Affect the Mind*, Edinburgh.

Benjamin Walter. [1963] (1998). *The Origin of German Tragic Drama* translated by J. Osborne, London: Verso.

Bent, Ian. (1980). "Analysis" in *The New Grove's Dictionary of Music and Musicians* vol. i, edited by S. Sadie, London: MacMillan: 340-388.

Berlin, Isaiah. [1979] (1981). *Against the Current: Essays in the History of Ideas*, Oxford: Oxford University Press.

Bernard, Jonathan W. (1980). "The Principles and the Elements: Rameau's Controversy with D'Alembert," *Journal of Music Theory* 24: 37-62.

Bernhard, Christoph. (1973). "The Treatises of Christoph Bernhard," translated and edited by W. Hilse, *The Musical Forum* 3: 1-196.

Black, Max. (1962). *Models and Metaphors*, Ithaca: Cornell University Press.

Blackall, Eric A. (1959). *The Emergence of German as a Literary Language 1700-1775*, Cambridge: Cambridge University Press.

Blackburn, Bonnie. (1987). "On Compositional Process in the Fifteenth Century," *Journal of the American Musicological Society* 40: 210-184.

Bloch, Ernst. (1985). *Essays on the Philosophy of Music*, translated by P. Palmer, Cambridge: Cambridge University Press.

Bond, Mark Evan. (1991). *Wordless Rhetoric: Musical Form and the Metaphor of Oration*, Cambridge. Mass.: Harvard University Press.

Booth, Wayne C. (1961). *The Rhetoric of Fiction*, Chicago: The University of Chicago Press.

Borgerhoff, E.B.O. (1950). *The Freedom of French Classicism*, Princeton: Princeton University Press.

Boyé, Pascal Boyer. [1779] (1973) *L'expression musicale mise au rang des chimères*, Amsterdam., Geneva: Minkoff.

Bremond, Claude. (1970). "Morphology of the French Folktale," *Semiotica* 2: 247-76.

Brook, Barry S. (1975). "Sturm und Drang and the Romantic Period in Music," *Studies in Romanticism* 9: 269-284.

Brown, Calvin S. (1948). *Music and Literature*, Athens: University of Georgia Press.

Brown, John. (The Estimate). (1763). *A Dissertation on the Rise, Union and Power, the Progressions, Separations and Corruptions of Poetry and Music*, London.

Brown, John. (The Painter). (1789). *Letters upon the Poetry and Music of Italian Opera*, London.

Broyles, Michael. (1980). "Organic Form and Binary Repeat," *Musical Quarterly* 66: 339-60.

Bruner, Jerome S. (1965). "Art as a Mode of Knowing," in *On Knowing*, Cambridge, Mass.: Harvard University Press: 59-74.

Bryson, Gladys. [1945] (1968). *Man and Society: The Scottish Inquiry of the Eighteenth Century*, New York: Augustus M. Kelly.

Bryson, Norman. (1981). *Word and Image: French Painting of the Ancien Régime*, Cambridge: Cambridge University Press.

Buelow, George. (1980). "Rhetoric and Music," *The New Grove's Dictionary of Music and Musicians* vol. xv, edited by by S. Sadie, London: MacMillan: 800-802.

________. (1983). "Johann Mattheson and the Invention of the *Affektenlehre*," *New Mattheson Studies*, edited by G. Buelow and H. Marx, Cambridge: Cambridge University Press: 393-407.

Bukofzer, Manfred F. (1939-40). "Allegory in Baroque Music," *The Journal of the Warburg Institute*, 3: 1-21.

________. (1948). *Music in the Baroque Era*, London: Dent.

Burke, Edmund. (1958). *A Philosophical Enquiry into the Origin of Our Ideas of the Sublime and Beautiful,* edited by with intro. J.T. Boulton, Notre Dame, Indiana: University of Notre Dame Press.

Burmeister, Joachim. [1601; 1601] (1993). *Musical Poetics*, translated with int. and notes, B. V. Rivera, edited by C. V. Palisca, New Haven: Yale University Press.

Bush, Douglas. (1961). "Science and Literature," *Seventeenth-Century Science and the Arts*, edited by H.H.
Rhys, Princeton: Princeton University Press: 29-62.

Butler, Gregory G. (1977). "Fugue and Rhetoric," *Journal of Music Theory* 21: 49-109.

________. (1980). "Music and Rhetoric in Early Seventeenth-Century English Sources" *Musical Quarterly* 66: 53-64.

Cassirer, Ernst. [1932] (1951). *The Philosophy of the Enlightenment*, translated by F. Köln and J. P. Pettegrove, Princeton: Beacon Press.

________. (1955). *The Philosophy of Symbolic Forms* in 3 vols. translated by R. Manheim, New Haven: Yale University Press.

Caygill, Howard. (1989). *Art of Judgement,* Oxford: Basil Blackwell.

Chabanon, Michel-Paul Guy de. (1785). *De la musique considéré en elle même et dans ses rapports avec la parole, les langues, la poêsie et la théâtre*, Paris.

Chafe, Eric. (1991). *Tonal Allegories in the Vocal Music of J.S. Bach*, Berkeley: University of California Press.

Chapman, H. Perry. (1990). *Rembrandt's Self-Portraits: A Study in Seventeenth-Century Identity,* Princeton: Princeton University Press.

Chastellux, Françios, Jean de. (1765). *Essai surl'union de la poésie et de las musique*, Paris.

Chomsky, Noam. (1957). *Syntactic Structures*, The Hague: Mouton.

Christensen, Thomas. (1985). *Science and Music Theory in the Enlightenment: D'Alembert's Critique of Rameau,* an unpublished Ph.D. dissertation, Yale University.

________. (1989). "Music Theory as Scientific Propaganda: The Case of D'Alembert's *Elemens de Musique*," *Journal of the History of Ideas* 50: 409-427.

Churgin, Bathia. (1963). *The Symphonies of G.B. Sammartini* in 2 vols. an unpublished Ph.D. dissertation, Harvard University.

Chytry, Josef. (1989). *The Aesthetic State: A Quest in Modern German Thought*, Berkeley: University of California State.

Clynes, Manfred. (1980). "The communication of emotion: theory of sentics" *Emotion: Theory, Research, and Experience,* vol. i, edited by R. Plutchik and H. Kellerman, Orlando: Academic Press: 271-301.

Cohen, Albert. (1981). *Music in the French Royal Academy of Science*, Princeton: Princeton University Press.

Cohen, H.F. (1984). *Quantifying Music: The Science of Music at the First Stage of the Scientific Revolution, 1580-1650*, Dodrecht: Reidel Publishing.

Condillac, Etienne Bonnot de. [1756] (1974). *An Essay on the Origin of Human Knowledge*, translated Th. Nugent, New York: AMS Press.

Cone, Edward T. (1974). *The Composer's Voice*, Berkeley: The University of California Press.

________. (1972). "Music: A View from Delft," *Perspectives in Contemporary Music Theory*, ed, B. Borez and E. T. Cone, New York: Norton.

________. (1982). "Schubert's Promissory Note: An Exercise in Musical Hermeneutics," *19th Century Music* 5: 233-241.

Cooke, Deryck. (1959). *The Language of Music*, London: Oxford University Press.

Cooper, John Gilbert. (1745). *The Power of Harmony*, London.

Costa, Gustano. (1976). "Vico's Influence on Eighteenth Century European Culture: A Footnote to Professor Nisbet's Paper," *Vico and Contemporary Thought*, edited by G. Tagliacozzo, M. Mooney and D.P. Verene, Atlantic Highlands, N.J.: Humanities Press: 247-9.

Couturat, L. (1961). *Opuscules et Fragments inedits de Leibniz*, Hildesheim: George Olms.

Cox, R. G. (1956). "A Survey of Literature from Donne to Marvell," *From Donne to Marvell* (*The Pelican Guide to English Literature,* vol iii), Baltimore: Penguin Books.

Crane, R. S. (1967). "The Plot of Tom Jones," *Twentieth Century Interpretations of Tom Jones,* edited by M.C. Ballesin, New York: Prentice Hall.

Croll, M. W. (1926). "The Baroque Style in Prose," *Studies in English Philology*, edited by E. Melon and M. B. Ruud, Minneapolis: 427-56.

Cropper, Elizabeth and Dempst, Charles. (1996). *Nicolas Poussin: Friendship and the Love of Painting*, Princeton University Press.

Culler, Jonathan. (1975). *Structuralist Poetics: Structuralism, Linguistics and the Study of Literature*, London: Routledge & Kegan Paul.

Dahlhaus, Carl. (1978). *Die Idee der Absoluten Musik*, Kassel: Barenreiter Verlag; translated as (1989) *The Idea of Absolute Music* translated R. Lustig, Chicago University Press.

Dahlhaus, Carl. [1968] (1990). *Studies on the Origin of Harmonic Tonality*, translated by Robert O. Gjerdingen, Princeton: Princeton University Press.

________. [1967] (1982). *Esthetics of Music*, translated W. Austin, Cambridge: Cambridge University Press.

________. (1975). "Fragmente zur musikalischen Hermeneutik," *Beiträge zur Musikalischen Hermeneutik*, edited by C. Dahlhaus, Regensburg: Gustav Boss Verlag.

________. (1980). "Counterpoint," *The New Grove's Dictionary of Music and Musicians* vol iv, edited by S. Sadie, London: 843-852.

Dammann, Rolf. (1967). *Der Musikbegriff im deutschen Barock*, Köln: Arno Volk Verlag.

Daniel, Samuel. [1930] (1950). *Poems and a Defence of Ryme* , edited by A. Sprague, London: Routledge and Kegan Paul.

Danto, Arthur C. (1981). *The Transfiguration of the Commonplace*, Cambridge, Mass.: Harvard University Press.

Darenberg, Karl H. (1960). *Studien zur Englischen Musikaesthetik des 18. Jahrhunderts*, Hamburg: Cram, de Gruyter.
David Hans T. and Mendal, Arthur. [1945] (1966). *The Bach Reader*, New York: Norton.
Davidson Reid, Jane (ed.). (1933). *The Oxford Guide to Classical Mythology in the Arts, 1300-1990s*, with the assistance of Chrid Rohmann, in 2 Vols. Oxford: Oxford University Press.
Debus, Allen G. (1979). *Robert Fludd and his Philosophical Key*, New York: Science History Publications.
Dent, Cathy H. (1987). "Development Studies of Perception and Metaphor: The Twain Shall Meet," *Metaphor and Symbolic Activity* 2: 53-71.
Derrida, Jacques. [1967] (1977). *Of Grammatology*, Baltimore: The Johns Hopkins University Press.
Descartes, Rene. [1618] (1961). *Compendium of Music*, translated by W. Robert, New York: The American Institute of Musicology.
________. (1985). *The Philosophical Writings of Descartes*, Vol. 1 translated by J. Cottingham, R. Stoothoff, and D. Murdoch, Cambridge: Cambridge University Press.
Dieckman, Liselotte. (1970). *Hieroglyphics*, St. Louis: Washington University Press.
Donington, Robert. (1963). *Wagner's Ring and Its Symbols: The Symbols and the Myth*, London: Faber & Faber.
________. (1968). "Monteverdi's First Opera," *The Monteverdi Companion*, edited by D. Arnold and N. Fortune, London: Faber & Faber.
Donne, John. (1952). *The Complete Poetry and Selected Prose of John Donne*, edited with and introduction by Charles Coffin, New York: The Modern Library.
Dorfles, Gilo. (1976). "Myth and Metaphor in Vico and in Contemporary Aesthetics," in *Vico and Contemporary Thought*, edited by G. Tagliacozzo, M. Mooney and D.P. Verene, Atlantic Highlands: Humanities Press: 557-590.
Dowling, Jay W. (1978). "Scale and Contour: Two Components of a Theory of Memory for Melodies," *Psychological Review* 85: 341-354.
________. (1982). "Melodic Information Processing and Its Development," in *The Psychology of Music*, edited by D. Deutsch, Orlando: Academic Press: 413-429.
Dowling, Jay W. and Harwood, Dane L. (1986). *Music Cognition*, Orlando: Academic Press.
Draper, John W. (1932-33). "Poetry and Music in Eighteenth-Century Aesthetics," *Englische Studien* 67: 70-85.
Dubos, Abbes. [1719]. *Réflexions critiques sur la poésie et la peinture* in 3 vols. Paris. translated as [1748] (1978). *Critical Reflections*, translated Th. Nugent, New York: AMS Press.
Eco, Umberto. (1976). *A Theory of Semiotics,* Bloomington: Indiana University Press.
________. (1981). *The Role of the Reader*, Bloomington: Indiana University Press.

________. (1984). *Semiotics and the Philosophy of Language*, Bloomington: Indiana University Press.

Engell, James. (1981). *The Creative Imagination: Enlightenment to Romanticism*, Cambridge, Mass.: Harvard University Press.

Fainsilber, Lynn and Ortony, Andrew. (1987). "Metaphorical Uses in the Expression of Emotions," *Metaphor and Symbolic Activity* 2: 239-250.

Fellows, Otis. (1977). *Diderot*, Boston: Twayne Publishers.

Fenlon, Ian. (1980). "Education in Music: Renaissance and Reformation Schools," *The New Grove's Dictionary of Music and Musicians,* vol. vi, edited by S. Sadie, London: Macmillan: 9-11.

________. (1985). "The Mantuan Stage Works," *The New Monteverdi Companion*, edited by D. Arnold and N. Fortune, London: Faber & Faber: 251-287.

Finney, Gretchen L. (1962). *Musical Backgrounds for English Literature*: 1580-1650, New Brunswick: Rutgers University Press.

Flage, Daniel E. (1990). *David Hume's Theory of Mind*, London: Routledge.

Flood, W.H. Grattan. (1916). "An Eighteenth Century Essayist on Poetry and Music," *Musical Quarterly* 2: 191-8.

Foucault, Michel. (1972). *The Order of Things: An Archaeology of the Human Sciences*, New York: Vintage Books.

Freedberg, David (1989). *The Power of Images: Studies in the History and Theory of Response,* Chicago: The University of Chicago Press.

Fraisse, Paul. (1982). "Rhythm and Tempo," in *The Psychology of Music*, edited by Diana Deutsch, Orlando: Academic Press: 149-180.

Franke, Ursula. (1975). "Von der Metaphysik zur Ästhetik: Der Schritt von Leibniz zu Baumgarten," *Studia Leibnizian, Akten des II internationalen Leibniz Kongresses* (Hannover 17-22 Juli 1972) 3: 229-240.

Fubini, Enrico. (1990). *The History of Music Aesthetics*, translated by M. Hatwell, London: Macmillan.

Gadamer, Hans-Georg. (1976). *Philosophical Hermeneutics,* translated and edited by David E. Linge, Berkeley: University of California Press.

Gardner, Howard. (1985). *The Mind's New Science: A History of the Cognitive Revolution*, New York: Basic Books.

Gardner, Howard and Winner, Ellen. (1979). "The Development of Metaphoric Competence: Implications for Humanistic Disciplines" *On Metaphor*, edited by Sheldon Sacks, Chicago: The University of Chicago Press: 121-130.

Gay, Peter. (1969). *The Enlightenment: An Interpretation* in 2 vols. New York: Alfred A. Knopf.

Genette, Gerard. [1972] (1980). *Narrative Discourse*, translated by J. Lewin, Ithaca: Cornell University Press.

Georgiades, Thrasysbulos. (1950). "Aus der Musiksprache des Mozarttheaters," *Mozart-Yahrbuch* 1: 76-98.

________. (1951). "Zur Musiksprache der Wiener Klassiker," *Mozart-Jahrbuch* 2: 50-59.

________. [1954] (1982). *Music and Language; The Rise of Western Music as Exemplified in the Setting of the Mass*, translated M. L. Gllener, Cambridge: Cambridge University Press.

Gewirtz, Allan. (1941). "Experience and the Non-Mathematical in the Cartesian Method," *Journal of the History of Ideas* 2: 183-210.

Godwin, Jocelyn. (1979). *Athanasius Kircher: A Renaissance Man and the Quest for Lost Knowledge*, London: Thames and Hudson.

Gombrich, E.H. (1957). "Lessing: Lectures on a Master Mind," *Proceedings of the British Academy,* 43: 133-56. Later (1984) published as "The diversity of Art: The Place of the *Laocoon* in the Life and Work of G. E. Lessing (1729-1781)" in *Tributes: Interpreters of Our Cultural Tradition*, Oxford: Phaidon Press.

________. [1960] (1977). *Art and Illusion: A Study in the Psychology of Pictorial Representation*, fifth edition, Oxford: Phaidon Press.

________. [1966] (1985). *Norm and Form: Studies in the art of the Renaissance I*, fourth edition, Oxford: Phaidon Press.

________. (1972). *Symbolic Images: Studies in the Art of The Renaissance II*, Oxford: Phaidon Press.

Goodman, Nelson. (1968). *Languages of Art*, Indianapolis: The Bobbs-Merrill Company.

________. (1978). *Ways of Worldmaking*, Indianapolis and Cambridge: Hackett Publishing Company.

Grant, Patrick. (1985). *Literature and the Discovery of Method in the English Renaissance*, London: Macmillan Press.

Greimas, Algirads J. [1969] (1977). "Elements of a Narrative Grammar," *Diacritics* 7: 23-40.

Grene, Marjorie. (1985). *Descartes*, Brighton: The Harvester Press.

Grierson, Herbert J. [1925] (1965). *Metaphysical Poetry from Donne to Butler*, Oxford: Clarendon Press.

Gruber, Albion. (1970). "Mersenne and Evolving Tonal Theory," *Journal of Music Theory* 14: 36-67.

HaCohen, Ruth (1998). "Generating Pragmatic Musical Grammar: The Case of Enlightened Scaffolding," in *Musical Signification Between Rhetoric and Pragmatics* (Proceedings of the 5th International Congress on Musical Signification), edited by by G. Stefani, E. Tarasti and L. Marconi, CLUEB,: 457-466.

________. (1999). "Planes and Their Interplay: The Alchemy of Forms and Emotions in St. Matthew Passion," in *Music and Sign: Semiotic and Cognitive Studies in Music*, edited by by Ioannis Zannos, 416-434 Bratislava: Asco Art and Sciences.

HaCohen, Ruth and Wagner, Naphtali. (1997). "The Communicative Force of Wagner's Leitmotifs: Complementary Relations between their Connotations and Denotations," *Music Perception* (14): 445-476.

Hagstrum, Jean H. (1958). *The Sister Arts: The Tradition of Literary Pictorialism in English Poetry from Dryden to Gray*, Chicago: The University of Chicago Press.

Hamilton, Kenneth G. (1963). *The Two Harmonies: Poetry and Prose in the Seventeenth Century*, Oxford: Clarendon Press.

Harrán, Don. (1986). *Word-Tone Relation in Musical Thought from Antiquity to the Seventeenth Century*, American Institute of Musicology, (Hansler Verlag).

Harris, James. [1744] (1792). *Three Treatises. The First Concerning Art. The Second Concerning Music, Painting and Poetry. The Third Concerning Happiness*, in *Miscellanies* vol 1. London: F.Wingrave.

Harris, John. (1977). "Leibniz and Locke on Innate Ideas," in *Locke on Human Understanding*, edited by I.C. Tipton, Oxford University Press.

Hatten, Robert. (1994). *Musical Meaning in Beethoven: Markedness, Correlation, Interpretation*, Bloomington: Indiana University Press.

Hausman, Carl R. (1989). *Metaphor and Art: Interactionism and Reference in the Verbal and Nonverbal Arts*, Cambridge: Cambridge University Press.

Hawkins, Sir John. (1776). *A General History of the Science and Practice of Music* in 5 vols. London.

Hecht, Hans. (1920). *Daniel Webb: Ein Beitrag zur englischen Aesthetik des achtzehnten Jahrhunderts*, Hamburg: Verlag von Henri Grand.

Heffernan, James A. W. (1985). *The Re-Creation of Landscape: A Study of Wordsworth, Coleridge, Constable, and Turner*, Hanover: University Press of New England.

Hell, Helmut. (1971). *Die Neapolitanische Opernsinfonie in der ersten Hälfte des 18 Jahrhunderts*, München: Hans Schneider Tutzing.

Heninger, Simeon K., Jr. (1978). *The Cosmographical Glass: Renaissance Diagrams of the Universe*, San Marino Calif.: The Huntington Library.

Herder, Johann Gottfried. (1966). "On the Origin of Language" *On the Origin of Language*, translated and edited by J.H. Moran and A. Gode, New York: Frederick Ungar: 87-176.

Hobbes, Thomas. [1651] (1968). *Leviathan*, edited by C.B. Macpherson, Penguin Books.

Hoffman, Paul. (1986). "The Unity of Descartes' Man," *Philosophical Review* 95: 339-392.

Hogan Michael J. (1984). "Historiography and Ethics in Adam Smith's Lectures on Rhetoric, 1762-1763," *Rhetorica* 2: 75-91.

Hollander, John. (1956). "The Music of Poetry," *Journal of Aesthetic and Art Criticism* 15: 232-244.

________. (1957). "Musica Mundana and Twelfth Night," *Sound and Poetry*, edited by N. Frye, New York: Columbia University Press: 55-82.

________. (1961). *The Untuning of the Sky*, New York: Norton.

Hosler, Bellamy. (1978). *Changing Aesthetic Views of Instrumental Music in 18th-Century Germany*, Ann Arbor: UMI Research Press.

Howard, William G. (1909). "Ut pictura poesis," *Publications of the Modern Language Association* 24: 40-123.

Hume, David. (1739). *A Treatise of Human Nature,* London.

Hunter, Paul J. (1975). *Occasional Form,* Baltimore: The John Hopkins University Press.

Hutcheson, Francis. (1973). *Inquiry Concerning Beauty, Harmony and Design*, edited by P. Kivy, The Hague: Martinus Nijhoff.

Ishiguro, Hidé. (1972). *Leibniz's Philosophy of Logic and Language*, London: Duckworth.

Izard, Carrol E. (1971). *The Face of Emotion*, New York: Appleton-Century-Crofts.

Jacob, Hilderbrand. (1734). *Of the Sister Arts, an Essay*, London.

Johnson, Paula. (1972). *Form and Transformation in Music and Poetry of the English Renaissance*, New Haven: Yale University Press.

Jones, Sir William. [1772] [1810] (1971). "On the Arts, Commonly Called Imitative," *The Works of the English Poets* vol. xviii, edited by A. Chalmers, Hildesheim: Georg Olms Verlag: 501-511.

Jones, William of Nayland. (1784). *A Treatise on the Art of Music*, Colchester.

Pauli, Wolfgang. (1955). "The Influence of Archetypal Ideas on the Scientific Theories of Kepler," in *TheInterpretation of Nature and Psyche,* edited by C. G. Jung and W. Pauli, New York: Pantheon Books.

Kallich, Martin. (1970). *The Association of Ideas and Critical Theory in Eighteenth-Century England*, The Hague.

Kant, Emanuel. [1790] (1952) *Critique of Judgement,* translated J. H. Meredith, Oxford: Clarendon Press.

Katz , Ruth. (1985). "Societal Codes for Responding to Dissent," in *The Planning of Change,* 4th edition, edited by W. Bennis et. al. New York: Holt, Rinehart and Winston.

________. (1986). *Divining the Powers of Music: Aesthetic Theory and the Origins of Opera*, New York: Pendragon Press.

________. (1991). "History as Compliance: The Development of Western Musical Notation in the Light of Goodman's Requirements," in *How Classifications Work*, edited by M. Douglas and D. Hull, Edinburgh: Edinburgh University Press:

Katz, Ruth and Dahlhaus, Carl. (1987-1991). *Contemplating Music: Source Readings in the Aesthetics of Music* in 4 vols. Vol. i: *Substance*; vol ii: *Import*; vol. iii: *Essence*; vol. iv: *Community of Discourse*. New York: Pendragon Press.

Katz, Ruth and HaCohen, Ruth (1988). "`*Ut Musica Poesis*': The Crystallization of a Conception Concerning Cognitive Processes and `Well-Made Worlds'," in *Festschrift Carl Dahlhaus*, edited by H. Danuser et al. Laaber-Verlag: Laaber: 17-37.

Kemp, Matin, ed. (1989). *Leonardo on Painting*, New Haven: Yale University Presss.

Kerman, Joseph. (1952). *Opera as Drama*, New York: Vintage Books.

King, Hyatt A. (1955). *Mozart in Retrospect,* Oxford: Oxford University Press.

Kirkendale, Ursula. (1980). "The Source for Bach's *Musical Offering*: The *Institutio Oratoria* of Quintilian," *Journal of the American Musicological Society* 31: 88-141.

Kirkendale, Warren. (1979). *Fuge and Fugato in Rococo and Classical Chamber Music*, Durham: Duke University Press.

Kivy, Peter. (1976). *The Seventh Sense,* New York: Burt Franklin.

________. (1980). *The Corded Shell: Reflections on Musical Expression*, Princeton: Princeton University Press.

________. (1984). *Sound and Semblance: Reflections on Musical Representation*, Princeton: Princeton University Press.

________. (1988). *Osmin's Rage: Philosophical Reflections on Opera, Drama, and Text*, Princeton: Princeton University Press.

________. (1989). *Sound Sentiment*: Princeton: Princeton University Press.

Koyré, Alexander. (1961). *From the Closed World to the Open Universe*, Baltimore: The Johns Hopkins University Press.

Kramer, Lawrence. (1984). *Music and Poetry: The Nineteenth Century and After*, Berkeley: University of California Press.

________. (1995). *Classical Music and Postmodern Knowledge*, Berkeley: University of California Press.

Kretzmann, Norman. (1977). "The Main Thesis of Locke's Semantic Theory," *Locke on Human Understanding*, edited by I.C. Tipton, Oxford: Oxford University Press.

Kristeller, Paul O. (1965). "The Modern System of the Arts," *Renaissance Thought and the Arts*, Princeton: Princeton University Press: 163-227.

Kuhn, Thomas S. (1962). *The Structure of Scientific Revolutions*, Chicago: The University of Chicago Press.

Lang, Paul H. (1942). *Music in Western Civilization*, London: Dent & Sons Ltd.

Langer, Suzanne K. (1953). *Feeling and Form*, New York: Charles Scribner and Sons.

________. [1942] (1976). *Philosophy in a New Key*, Cambridge, Mass.: Harvard University Press.

________. (1957). "The Art Symbol and the Symbol in Art," in *Problems of Art*, New York: Charles Scribner and Sons.

________. [1958] (1961). *Reflections on Art*: New York: Oxford University Press.

Larsen, J. P. (1967). "Some Observations on the Development and Characteristics of Vienna Classical Instrumental Music," *Studia Musicologica* 9: 115-139.

Lee, Rensselaer W. [1940] (1967). *Ut pictura poesis: The Humanistic Theory of Painting*, New York: Norton.

Le-Hurray, Peter and Day, James, eds. (1981). *Music and Aesthetics in the Eighteenth and Early Nineteenth Centuries*, Cambridge: Cambridge University Press.

Leibniz, Gottfried Wilhelm, Freiherr von. (1969). *Philosophical Papers and Letters*, edited by L. E. Loemker, Dordrecht: Reidel.

Lerdahl, Fred, and Jackendoff, Ray. (1983). *A Generative Theory of Tonal Music*, Cambridge, Mass.: MIT Press.

Lessing, Gotthold, Ephraim. [1766] (1957). *Laocoon: An Essay Upon the Limits of Painting and Poetry,* translated E. Fothinham, New York: The Noonday Press.

________. [1768-9] (1958). *Hamburgische Dramaturgie*, Stuttgart: Alfred Kröner Verlag.

Lester, Joel. (1992) *Compositional Theories* , Cambridge, Mass.: Harvard University Press.

Levi-Strauss, Claude. [1964] (1969). *The Raw and the Cooked*, translated J. D. Weightman, London: Jonathan Cape.

________. [1963] (1977). *Structural Anthropology*, Translated C. Jacobson, New York: Penguin Books.

Lippman, Edward D. (1964). *Musical Thought in Ancient Greece*, New York: Columbia University Press.

Lipking, Lawrence. (1970). *The Ordering of the Arts in Eighteenth-Century England*, Princeton, N.J.: Princeton University Press.

________. ed. (1986). *Musical Aesthetics: A Historical Reader* in 2 vols. New York: Pendragon Press.

Livingston, Herbert S. (1952). *The Italian Overtures from Alessandro Scarlatti to Mozart*, an unpublished Ph.D. Dissertation, The University of North Carolina.

Locke, Arthur W. (1935). "Descartes and Seventeenth Century Music," *Musical Quarterly* 1: 423-431.

Locke, John. [1690] (1961). *An Essay Concerning Human Understanding*, London: Dent.

Lovejoy, Arthur O. (1960). *Essays in the History of Ideas*, Baltimore: The Johns Hopkins University Press: 69-77.

________. (1933). "Monboddo and Rousseau," *Modern Philology* 3: 275-296.

________. [1936] (1976). *The Great Chain of Being*, Cambridge, Mass.: Harvard University Press.

________. (1904). "Some Eighteenth Century Evolutionists," *Popular Science Monthly* 65: 336-40.

Lowinsky, Edward E. (1965). "Taste, Style and Ideology in Eighteenth-Century Music," *Aspects of the Eighteenth Century*, edited by E.R. Wasserman, Baltimore: The Johns Hopkins University Press: 163-205.

McCormac, Earl R. (1985). *A Cognitive Theory of Metaphor*, Harvard: MIT Press.

________. (1986). "Creative Metaphors," *Metaphor and Symbolic Activity* 1: 171-184.

Mace, Dean T.(1952). *English Musical Thought in the Seventeenth Century: A Study of an Art in Decline*, an unpublished Ph.D. dissertation, Columbia University.

________. (1970). "Marin Mersenne on Language and Music," *Journal of Music Theory* 14: 3-35.

Man, Paul de. [1971] (1983). "The Rhetoric of Blindness: Jacques Derrida's Reading of Rousseau," in *Blindness and Insight, Essays in the Rhetoric of Contemporary Criticism,* 2nd ed., Minneapolis: University of Minnesota Press: 102-141.

Maniates, M.R. (1969). "'Sonate, que me veux-tu': The Enigma of French Musical Aesthetics in the Eighteenth Century," *Current Musicology* 9: 117-40.

Manwaring, Edward. (1744). *Of Harmony and Numbers in Latin and English Prose and in English Poetry*, London.

Markiewicz, J. (1972). "*Ut Pictura Poesis*," *Journal of Aesthetic and Art Criticism* 31.

Mason, John. (1749). *An Essay on the Power of Numbers and the Principles of Harmony in Poetical Compositions*, London.

Mattheson, Johann. (1739). *Der Volkommene Cappelmeister*, Hamburg: Christian Herald.

translated as (1981). *Der Volkommene Cappelmeister*, by E. Harriss, Ann Arbor: UMIResearch Press.

Mendelssohn, Moses. [1929] (1971-2). *Gesammelte Schriften* vol. i-ii, Stuttgart: Friedrich Frommann Verlag.

Mersenne, Marin. [1636] (1975). *Harmonie Universelle*, Facs. ed by the Centre National de la Recherche Scientifique.

Meyer, Leonard B. (1956). *Emotions and Meaning in Music*, Chicago: University of Chicago Press.

________. (1980). "Exploiting Limits: Creation, Archetypes and Style Change," *Daedalus* 109: 177-205.

Miall, David S. (1987). "Metaphor and Affect: The Problem of Creative Thought," *Metaphor and Symbolic Activity* 2: 81-96.

Michael, Emily. (1984). "Francis Hutcheson on Aesthetic Perception and Aesthetic Pleasure," *British Journal of Aesthetics* 24: 242-255.

Milton, John. (1957). *Complete Poems and Major Prose*, edited by M.Y. Hughes, Indianapolis: The Odyssey Press.

Mitford, William. (1774). *Essay upon the Harmony of Language*, London.

Monboddo, James Burnett, Lord. [1773-92] (1809). *On the Origin and Progress of Language* in 5 vols. Edinburgh: A. Smellie.

Mooney, Michael.(1985). *Vico in the Tradition of Rhetoric*, Princeton: Princeton University Press.

Mora, George. (1976). "Vico, Piaget and Genetic Epistemology," *Giambattista Vico's Science of Humanity*, edited by G. Tagliacozzo and D.P. Verene, Baltimore: The Johns Hopkins University Press: 341-350.

Murphy, James. (1974). *Rhetoric in the Middle Ages: A History of Rhetorical Theory from Saint Augustine to the Renaissance*, Berkeley: University of California Press.

Nagel, Ernst. [1961] (1982). *The Structure of Science: Problems in the Logic of Scientific Explanation*, London: Routledge & Kegan Paul.

Narmour, Eugene. (1992). *The Analysis and Cognition of Melodic Complexity: the Implication-Realization Model*, Chicago: University of Chicago Press.

Neubauer, John (1986). *The Emancipation of Music from Language*, New Haven: Yale University Press.

Newman, John. (1963). *The Sonata in the Classic Era*, Chapel Hill: The University of North Carolina Press.

Newcomb, Anthony (1984). "Once More `Between Absolute and Program Music': Schumann Second Symphony," *19th Century Music* 7: 233-50.

________. (1987). "Schumann and Late Eighteenth Century Narrative Strategies," *19th Century Music* 11: 164-174.

Nicolson, Marjorie. (1960). *The Breaking of the Circle: Studies in the Effect of the 'New Science' upon Seventeenth-Century Poetry*, New York: Columbia University Press.

Niesser, Ulrich. (1967). *Cognitive Psychology*, New York: Appleton-Century-Crofts.

Nietzsche, Friedrich. (1968). *Basic Writings of Nietzsche* vol. i, translated and edited by W. Kaufmann, New York: The Modern Library.

Nisbet, H. B., ed. (1985). *German aesthetic and Literary Criticism: Winckelmann, Lessing, Hamann, Herder, Schiller, Goethe*. Cambridge: Cambridge University Press.

Oliver, Alfred. (1947). *The Encyclopédistes as Critics of Music*, New York: Columbia University Press.

Omond, T.S. (1907). *English Metrists*, London: Oxford University Press.

Oras, Ants. (1956). "Spenser and Milton: Some Parallels and Contrasts in the Handling of Sound,"*Sound and Poetry*, edited by N. Frye, New York: Columbia University Press: 109-133.

Orsini, Giordano. (1972). "The Ancient Roots of a Modern Idea," *Organic Form, The Life of an Idea*, edited by G.S. Rousseau, London: Routledge and Kegan Paul: 25-60.

Osgood, C. E., Suci G. J. & Tannenbaum P. H. (1957). *The Measurement of Meaning*, Urbana: University of Illinois Press.

Palisca, Claude V. (1960). "Vincenzo Galilei and Some Links between Pseudo Monody and Monody," *Musical Quarterly* 46 : 344-60.

________. (1961). "Scientific Empiricism in Musical Thought," *Seventeenth Century Science and the Arts*, edited by H. H. Rhys, Princeton: Princeton University Press, 1961: 91-137.

________. (1972). "Ut oratoria musica: The Rhetorical Basis of Musical Mannerism," *The Meaning of Mannerism*, edited by F.W. Robertson and S.G. Nichols, Hanover, University of New England: 37-65.

________. [1968] (1981). *Baroque Music*, New York: Prentice Hall.

________. (1981a). "Peri and the Theory of Recitative," *Studies in Music* 15: 51-61.

________. (1985). *Humanism in Italian Renaissance Musical Thought*, New Haven: Yale University Press.

Parks, Richard S. (1984). *18th-Century Counterpoint and Tonal Structures*, New York: Prentice Hall, 1984.

Panofsky, Erwin. [1924] (1968). *Idea, A Concept in Art Theory*, translated by J. S. Peake, New York: Harper and Row.

________. [1955] (1983). *Meaning in the Visual Arts*, Penguin Books.

Passmore, John. (1972). *The Perfectability of Man*, London, Duckworth.

Pater, Walter. [1893] (1980). "The School of Giorgione," *The Renaissance*, the 1893 Text, edited by D.L. Hill, Berkeley: University of California Press: 102-122.

Patey, Douglas L. (1984). *Probability and Literary Form*, Cambridge: Cambridge University Press.

Patrides, ed. (1970). *The Cambridge Platonists*, Cambridge, Massachusetts: Harvard University Press.

Pevsner, Nikolaus. (1940). *Academies of Art Past and Present,* London: Cambridge University Press.

Pirro, Andre. (1907). *L'esthétique de Bach*, Paris: Fischbacher.

________. [1907] (1973). *Descartes et la Musique*, Geneve: Minkoff Reprint.

Pirrota, Nino and Povoledo, Elena. [1969] (1982). *Music and Theater from Poliziano to Monteverdi*, Cambridge: Cambridge University Press.

Plutchik, Robert. (1980) "A General Psychoevolutionary Theory of Emotion" *Emotion: Theory, Research and Experience* vol. i (Theories of Emotions), edited by R. Plutchik and H. Kellerman, Orlando: Academic Press.

Polanyi, Michael. (1966). *The Tacit Dimension,* Garden City: Doubleday.

The Polite Arts, or, a Dissertation on Poetry, Painting, Musick, Architecture and Eloquence (1749 anon.) London.

Pons, Alain. (1969). "Vico and French Thought," *Giambattista Vico: An International Symposium*, ed G. Tagliacozzo with H. Y. White, Baltimore: The Johns Hopkins University Press: 165-185.

Pope, Alexander. [1869] (1911). *The Poetical Works*, London: Macmillan.

Popkin Richard H. (1960). *The History of Skepticism Erasmus to Descartes*, Assen: Var Gorcum.

________. (1967). "Mersenne, Marin," *The Encyclopedia of Philosophy,* vol. v, edited by P. Edwards, London: Macmillan: 282-3.

Powicke, Fredrick J. [1926] (1971). *The Cambridge Platonists: A Study*, Hamden, Conn.: Archon Books.

Powers, Harold. (1980). "Language Models and Musical Analysis," *Ethnomusicology* 24: 1-60.

Pribram, Karl H. (1982). "Brain Mechanism in Music: Prolegomena for a Theory of the Meaning of Meaning," *Music, Mind and Brain*, edited by M. Clynes, New York: Plenum: 21-35.

Propp, Vladimir. [1928] (1968). *Morphology of the Folktale*, Austin: University of Texas Press.

Quintilianus, Aristides. (1983). *On Music* in 3 vols. edited by T. J. Mathiesen, New Haven: Yale University Press.

Rameau, Jean-Philippe. [1722] (1971). *Treatise on Harmony*, translated and edited by Ph. Gossett, New York: Dover Publications.

Ratner, Leonard G. (1956). "Eighteenth-Century Theories of Musical Period Structure," *Musical Quarterly* 42: 439-454.

________. (1970). "Ars Combinatoria: Chance and Choice in Eighteenth-Century Music," in *Studies in Eighteenth-Century Music: A Tribute to Karl Geiringer*, edited by H. C. Robbins Landon, Oxford: Oxford University Press: 343-363.

________. (1980). *Classic Music, Expression, Form and Style*, New York: Scribner Books.

Richards, I. E. (1936). *The Philosophy of Rhetoric*, Oxford: Oxford University Press.

Ricoeur, Paul. [1978] (1986). *The Rule of Metaphor,* translated R. Czerny, Oxford: Routledge and Kegan Paul.

Ringer, Alexander (1980). "Education in Music: 1600-1800," *The New Grove's Dictionary of Music and Musicians* vol xi, edited by S. Sadie, Macmillan: London: 12-18.

Roberts, Donald Ramsay. (1947). "The Music of Milton," *Philological Quarterly* 26: 328-344.

Robertson, Thomas. (1784). *An Inquiry into the Fine Arts*, London.

Rogerson, Brewster. (1945). *'Ut Musica Poesis': The Parallel of Music and Poetry in Eighteenth Century Criticism*, an unpublished Ph. D. dissertation, Princeton University.

________. (1953). "The Art of Painting the Passions," *Journal of the History of Ideas* 14 :68-94.

Rosen, Charles. [1971] (1976). *The Classic Style: Haydn, Mozart, Beethoven*, London: Faber & Faber.

________. (1980). *Sonata Forms*, New York and London: Norton.

Rosner, Burton S. and Meyer, Leonard B. (1982). "Melodic Processes and the Perception of Music," *The Psychology of Music*, edited by D. Deutsch, Orlando: Academic Press: 317-343.

Ross, Stephanie. (1984). "Painting the Passions: Charles LeBrun's *Conference sur l'Expression*," *Journal of the History of Ideas* 45: 25-47.

Rousseau, Jean-Jacques. (1827). *Oeuvres complètes* vol. xv, Paris.

________. [1753] (1966). "On the Origin of Languages," *On the Origin of Language*, translated and edited by J.H. Moran and A. Gode, New York: Fredrick Ungar: 3-83.

Rumelhart, David. (1975). "Notes on a schema for stories," *Representation and Understanding: Studies in Cognitive Science.* edited by D. Bobrow & A. Collins, Orlando: Academic Press:

Sacks, Sheldon, [ed.] (1979). *On Metaphor*, Chicago: The University of Chicago Press.

Saintsbury, George. (1923). *A History of English Prosody from the Twelfth Century to the Present Day*, in 3 vols. London: Macmillan.

Sartre, Jean Paul. (1950). *What is Literature*, translated B. Freshtman, London: Methuen.

Schrade, Leo. (1964). *Tragedy and the Art of Music,* Cambridge Mass.: Harvard University Press.

Schroeder, David P. (1990). *Haydn and the Enlightenment: The Late Symphonies and their Audience*, Oxford: Oxford Clarendon Press.

Schueller, Herbert M. (1948). "'Imitation' and 'Expression' in British Music Criticism in the Eighteenth Century," *Musical Quarterly* 34: 544-566.

________. (1947). "Literature and Music as Sister Arts: An Aspect of Aesthetic Theory in Eighteenth-Century Britain," *Philological Quarterly* 26 :193-205.

________. (1953). "Correspondences between Music and the Sister Arts, According to 18th Century Aesthetic Theory," *Journal of Aesthetics and Art Criticism* 11: 334-359.

Schweitzer, Albert. [1908] (1975). *Johann Sebastian Bach* in 2 vols. translated E. Newman, New York: Dover.

Shaftesbury, Anthony Ashley Cooper, Earl of. (1743). *Characteristics* in 3 vols. London.

Shapiro, Michael. (1983). *The Sense of Grammar.* Bloomington: Indian University Press.

Sheldon, David A. (1975). "The Galant Style Revisited and Re-evaluated," *Acta Musicolgica* 46: 240-269.

Shen, Yeshayahu (1989). "The X-Bar Grammar for Stories: Story Grammar revisited," *Text* 9: 415-467.

Shirlaw, Matthew. [1917] (1969). *The Theory of Harmony*, New York: Da Capo Press.

Smallwood, P. J., ed. (1989). *The Johnson Quotation Book: Based on the Collection of Chartres Bryson*, Bristol: Bristol Classical.

Smith, Adam. (1980). *Essays on Philosophical Subjects*, edited by W.P.D. Wightman, et al., Oxford: Clarendon Press.

________. [1759] (1976). *The Theory of Moral Sentiments,* edited By D. D. Raphael and A.L. Macfie, Oxford: Clarendon Press.

Sondheimer, Robert. (1922). "Die Formale Entwicklung der Vorklassische Sinfonie," *Archive für Musikwissenschaft* 4: 85-99.

Spencer, John R. (1957). "*Ut Rhetorica Pictura*: A Study in Quatrocento Theory of Painting," *The Journal of Warburg and Courtauld Institutes* 20: 26-44.

Spitzer, Leo. (1963). *Classical and Christian Ideas of World Harmony*, Baltimore: The Johns Hopkins University Press.

Stam, James H. (1974). "An Introduction to the New Edition of *An Essay on the Origin of Human Knowledge*," in E. B. de Condillac *An Essay on the Origin of Human Knowledge*, New York: AMS Press: 1-10.

Starobinsky, Jean [1971] (1988). *Jean-Jacques Rousseau, Transparency and Obstruction,* translated by A. Goldhammer, Chicago: The University of Chicago Press.

Stevens, John. (1986). *Words and Music in the Middle Ages,* Cambridge: Cambridge University Press.

Street, Alan. (1987). "The Rhetorico-Musical Structure of the 'Goldberg' Variations: Bach's *Clavier Ubung IV* and the *Institutio oratoria* of Quintilian," *Music Analysis* 6: 89-127.

Strohm, Reinhard. (1997). *Drama per Musica: Italian Opera Seria of the Eighteenth Century*, New Haven: Yale University Press.

Strunk, Oliver. [1950] (1981). *Source Readings in Music History* in 5 vols. New York: Norton.

Summers, David. (1987). *The Judgement of Sense: Renaissance Naturalism and the Rise of Aesthetics*, Cambridge: Cambridge University Press.

Sundberg, Johan, (1982). "Speech, Song and Emotions," in *Music, Mind and Brain*, edited by M. Clynes, New York: Plenum: 137-149.

Toulmin, Stephen. (1961). "Seventeenth-Century Science and the Arts," *Seventeenth Century Science and the Arts*, edited by H.H. Rhys, Princeton: Princeton University Press.

Townsend Danbey. (1982). "Shaftesbury's Aesthetic Theory," *Journal of Aesthetic and Art Criticism*: 205-213.

Treitler, Leo. (1988). "Mozart and the Idea of Absolute Music," *Festschrift Carl Dahlhaus*, Laaber: Laaber Verlag: 413-440. Also in Treitler, L. (1989) *Music and the Historical Imagination* Cambridge Mass.: Harvard University Press: 176-214.

Tsur, Reuven. (1977). *A Perception-Oriented Theory of Meter*, Tel Aviv: The Porter Israeli Institute for Poetics and Semiotics.

________. (1998). *Poetic Rhythm: Structure and Performance, an Empirical Study in Cognitive Poetics,* Berne: P. Lang.

Twining, Thomas. (1789). *Two Dissertations: 1) On Poetry Considered as an Imitative Art, 2) On The Different Senses of the World Imitative as Applied to Music by the Antients and by the Moderns*, in Aristotle's *Treatise of Poetry,* Translated with Notes, London.

Ullman, Shimon. (1980). "Against Direct Perception," *Behavioral and Brain Sciences* 3: 373-416.

Unger, Hans-Heinrich. (1941). *Die Beziehungen zwischen Musik und Rhetorik im 16.-18. Jahrhundert,* Würzburg.

Van Ghent, Dorothy. [1953] (1961). *The English Novel - Form and Function*, New York: Holt, Rinehart and Wiston.

Verba, Cynthia. (1993). *Music and the French Enlightenment: Reconstruction of a Dialogue 1750-1764,* Oxford: Clarendon Press.

Verene, D. P. (1981). *Vico's Science of Imagination*, Ithaca: Cornell Universty Press.

Vickers, Brian. (1970). *Classical Rhetoric in English Poetry*, London: Macmillan.

________. (1984). "Figures of Rhetoric/Figures of Music?" *Rhetorica* 2: 1-44.

________. (1988). *In Defence of Rhetoric*, Oxford: Clarendon Press.

Vico, Giambattista.[1744] (1963). *La scienza nouva* in 2 vols, Milano: Rizzoti.

Voltaire, M. d. (1772). *The Dramatic Works of M. de Voltaire* vol. xxi, translated Rev. Mr. Franklin, Dublin: R. Marchbank.

Walker, D.P. (1941, 1942). "Musical Humanism in the 16th and 17th Centuries," *The Music Review* 2: 1-13; 111-121 221-228; 288-308; 3: 55-71.

________. (1958). *Spiritual and Demonic Magic from Ficino to Campanella*, London: Warburg Institute.

________. (1978). *Studies in Musical Science in the Late Renaissance*, London: Warburg Institute.

Warburton, William. (1738). *The Divine Legation of Moses Demonstrated from the Principles of a Deist*, in 2 vols. London.

Watkins, Anthony J. and Dyson, Mary C. (1985). "On the Perceptual Organization of Tone Sequences and Melodies," *Musical Structure and Cognition,* Orlando: Academic Press: 149-180.

Watt, Ian. (1957). *The Rise of the Novel*, Harmondsworth: Chatto & Windus.

Webb, Daniel. (1760). *An Inquiry into the Beauties of Painting*, London.

________. (1762). *Remarks on the Beauties of Poetry*, London.

________. (1769). *Observations on the Correspondence between Poetry and Music*, London.

Weber, Carl. (1921) (1958). *The Rational and Social Foundations of Music*, Southern Illinois University Press.

Weber, William. (1992). *The Rise of Musical Classics in Eighteen-Century England,* Oxford: Oxford University Press.

Webster, James. (1974). "Towards a History of Viennese Chamber Music in the Early Classical Period," *Journal of the American Musicological Society* 27: 212-247.

Weinberg, Bernard. (1961). *A History of Literary Criticism in the Italian Renaissance* in 2 vols. Chicago: The University of Chicago Press.

Weisstein, Ulrich. [1964] (1969). *The Essence of Opera*, New York: Norton.

Wellek, Rene. [1955] (1981). *A History of Modern Criticism 1750-1950,* vol. i, Cambridge: Cambridge University Press.

________. (1969). "The Supposed Influence of Vico on England and Scotland in the Eighteenth Century," *Giambattista Vico: An International Symposium*, ed. G. Tagliacozzo with H. Y. White, Baltimore: The Johns Hopkins University Press: 215-223.

Wiesenfarth, Joseph. (1975). "Emma: Point Counterpoint," *Jane Austin: Bicentenary Essays*, edited by John Halperin, Cambridge: Cambridge University Press.

Willheim, Imanuel. (1963). *J.A. Schiebe—German Musical Thought in Transition*, Ph.D. diss., University of Illinois.

Willy, Basil. [1940] (1961). *The Eighteenth Century Background*, Boston: Beacon Press.

Wilson, John. (1959). *Roger North on Music,* London: Novello.

Winn, James Anderson. (1981). *Unsuspected Eloquence: A History of the Relations between Poetry and Music*, New Haven: Yale University Press.

Wittgenstein, Ludwig. (1980). *Culture and Value,* translated Peter Winch, Oxford: Oxford University Press.

Wittkower, Rudolf. (1971). *Architectural Principles in the Age of Humanism*, New York: W. W. Norton.

Wolf, Eugene K. (1981). *The Symphonies of Johann Stamitz*, Utrecht: Shellema & Holkema.

Wortshorne, Simon T. (1954). *Venetian Opera in the Seventeenth Century*, Oxford: Clarendon Press.

Yates, Francis. (1982) "The Art of Ramon Lull," in *Collected Papers* vol. i, London: Routledge and Kegan Paul.

Yolton, John, W. (1985). *Locke - An Introduction,* Oxford: Basil Blackwell.

Zuckerkandl, Victor. [1956] (1973). *Sound and Symbol*, Princeton: Princeton University Press.

Name Index

Subject Index